Philosophies, Puzzles and Paradoxes

Unlike mathematics, statistics deals with real-world data and involves a higher degree of subjectivity due to the role of interpretation. Interpretation is shaped by context as well as the knowledge, preferences, assumptions and preconceptions of the interpreter, leading to a variety of interpretations of concepts as well as results. *Philosophies, Puzzles and Paradoxes: A Statistician's Search for Truth* thoroughly examines the distinct philosophical approaches to statistics – Bayesian, frequentist and likelihood – arising from different interpretations of probability and uncertainty. These differences are highlighted through numerous puzzles and paradoxes and illuminated by extensive discussions of the background philosophy of science.

Features

- Exploration of the philosophy of knowledge and truth and how they relate to deductive and inductive reasoning, and ultimately scientific and statistical thinking.
- Discussion of the philosophical theories of probability that are wider than the standard Bayesian and frequentist views.
- Exposition and examination of Savage's axioms as the basis of subjective probability and Bayesian statistics.
- Explanation of likelihood and likelihood-based inference, including the controversy surrounding the likelihood principle.
- Discussion of fiducial probability and its evolution to confidence procedure.
- Introduction of extended and hierarchical likelihood for random parameters, with the recognition of confidence as extended likelihood, leading to epistemic confidence as an objective measure of uncertainty for single events.
- Detailed analyses and new variations of classic paradoxes, such as the Monty Hall puzzle, the paradox of the ravens, the exchange paradox, and more.
- Substantive yet non-technical, catering to readers with only introductory exposure to the theory of probability and statistics.

This book primarily targets statisticians in general, including both undergraduate and graduate students, as well as researchers interested in the philosophical basis of probability and statistics. It is also suitable for philosophers of science and general readers intrigued by puzzles and paradoxes.

Yudi Pawitan graduated with a PhD in statistics in 1987 from the University of California at Davis and has been a professor of biostatistics since 2001 at the Department of Medical Epidemiology and Biostatistics, Karolinska Institutet, Stockholm, Sweden. He has worked in many areas of statistical applications, including time series analyses and medical imaging, and for the last 20 years in the modelling and analysis of high-throughput genetic and molecular data with applications in cancer. He has published more than 200 peer-reviewed research papers, split about equally between methodology and applied publications. He is the author of the monograph *In All Likelihood* (2001) and co-author of *Generalized Linear Models with Random Effects* (2017) together with Youngjo Lee and John Nelder, both covering likelihood-based statistical modelling and inference. Philosophy of science, statistical puzzles and paradoxes have been lifelong interests.

Youngjo Lee graduated with a PhD in statistics in 1983 from Iowa State University. He is currently a professor emeritus of statistics at Seoul National University, an endowed-chair professor of data and knowledge service engineering at Dankook University, and a vice president of the Korean Academy of Science and Technology. Alongside the late John Asworth Nelder, he is an originator of hierarchical generalized models and h-likelihood, having co-authored over 200 peer-reviewed research papers on the application of h-likelihood in various statistical areas. He is also a co-author of monographs on h-likelihood theory and applications. Furthermore, he has developed related software and is currently extending h-likelihood procedures to deep neural networks.

Philosophies, Puzzles and Paradoxes

A Statistician's Search for Truth

Yudi Pawitan and Youngjo Lee

CRC Press
Taylor & Francis Group
Boca Raton London New York

CRC Press is an imprint of the
Taylor & Francis Group, an **informa** business
A CHAPMAN & HALL BOOK

First edition published 2024
by CRC Press
2385 Executive Center Drive, Suite 320, Boca Raton, FL 33431, U.S.A.

and by CRC Press
4 Park Square, Milton Park, Abingdon, Oxon, OX14 4RN

CRC Press is an imprint of Taylor & Francis Group, LLC

Library of Congress Cataloging-in-Publication Data

Names: Pawitan, Yudi, author. | Lee, Youngjo, author.
Title: Philosophies, puzzles and paradoxes : a statistician's search for truth / Yudi Pawitan and Youngjo Lee.
Description: First edition. | Boca Raton : CRC Press, 2024. | Includes bibliographical references and index. |
Summary: "Mathematics is focused on formal manipulation of abstract concepts, while statistics deals with real-world data and involves a higher degree of subjectivity due to the role of interpretation. Interpretation is shaped by context as well as the knowledge, biases, assumptions or preconceptions of the interpreter, leading to a variety of potential interpretations of concepts as well as results. This book thoroughly examines the distinct philosophical approaches to statistics--Bayesian, frequentist, and likelihood--arising from different interpretations of probability and uncertainty. These differences are highlighted through a variety of puzzles and paradoxes"-- Provided by publisher.
Identifiers: LCCN 2023039429 (print) | LCCN 2023039430 (ebook) | ISBN 9781032377391 (paperback) | ISBN
9781032377407 (hardback) | ISBN 9781003341659 (ebook)
Subjects: LCSH: Probabilities--Philosophy. | Statistics--Philosophy. | Statistical decision. | Subjectivity.
Classification: LCC QA273.A35 P395 2024 (print) | LCC QA273.A35 (ebook) |
DDC 519.501--dc23/eng/20231221
LC record available at https://lccn.loc.gov/2023039429
LC ebook record available at https://lccn.loc.gov/2023039430

ISBN: 978-1-032-37740-7 (hbk)
ISBN: 978-1-032-37739-1 (pbk)
ISBN: 978-1-003-34165-9 (ebk)

DOI: 10.1201/9781003341659

Typeset in CMR10
by KnowledgeWorks Global Ltd.

Publisher's note: This book has been prepared from camera-ready copy provided by the authors.

Contents

Foreword

This is a remarkable book: wide-ranging, ambitious, challenging and profound but also intriguing, fascinating and original. Despite the fact that I have been thinking about the foundations of statistics for a long time, in particular as to how those foundations relate to what we want to do and achieve as practising statisticians, I learned a lot from reading it and also unlearned some things I thought I knew.

Yudi Pawitan and Youngjo Lee are well-respected statisticians who, sometimes working together but also apart, have both made major contributions to our understanding of the role of likelihood in statistical inference and to its continuing development as a statistical tool. The book they have now written together is a philosophical treatise based on their deep understanding of likelihood and related statistical concepts such as confidence, and they have extended the scope of these tools to address evidential paradoxes often treated in philosophy, such as, for example, Hempel's paradox of the ravens.

It was John Nelder who first drew my attention to the importance of Yudi's by now classic book on likelihood (1) and it was through my friendship with John that I got to know of the ground-breaking work Youngjo was doing with John in extending likelihood to analysing hierarchical generalised linear models (2). Eventually the three collaborated to produce a book that did for hierarchical models (3) what John's famous book (4) with Peter McCullagh did for generalised linear models. John died in 2010 but his two coauthors have continued their collaboration and this has now taken this surprising turn away from modelling itself to consider more foundational issues.

The book has four parts. Part I covers a view of epistemology from the perspective of the statistician. Statisticians unfamiliar with the philosophy of science will find it a valuable introduction to that field. Philosophers of science, whether or not they are in agreement with this survey of their field, should find it a useful introduction to how statisticians might think about it.

Part II addresses probability and inverse probability. The former is the province of mathematics. Even if there are disagreements as to what probability means (and the book presents the various theories), its development can proceed largely axiomatically. However, any student of probability must occasionally wonder how anyone would ever know what a probability could be. The inverse passage from evidence to probability that such knowledge would involve is in the domain of statistics and raises many difficulties, which are dealt with here.

Part III is a thorough survey of likelihood and likelihood-based inference, a field in which the authors, as already noted, are experts. In what was a surprise to me, confidence emerges as a fundamental related concept. (One can only wonder what R.A. Fisher would think about a term of Neyman's he despised being related to a concept such as likelihood, which he regarded as being very much his own!)

Part IV uses the understanding gained in the earlier part to probe and resolve various notorious paradoxes. Quite apart from anything else, this section is very great fun, providing new insights on old favourites such as the Monty Hall Problem.

I very much enjoyed reading this highly original work and recommend it to all who are interested in statistical inference, evidence and the philosophy of science.

Stephen Senn
Edinburgh

References

1. Pawitan, Y. (2001). *In All Likelihood: Statistical Modelling and Inference Using Likelihood.* Oxford: Clarendon Press.

2. Lee, Y. and Nelder, J.A. (1996). Hierarchical generalized linear models (with discussion). *Journal of the Royal Statistical Society: Series B (Methodology)*, **58**, 619–678.

3. Lee, Y., Nelder, J.A. and Pawitan, Y. (2006). *Generalized Linear Models with Random Effects: Unified Analysis via H-likelihood.* 1st Edition. Boca Raton: Chapman & Hall/CRC.

4. McCullagh, P. and Nelder, J.A. (1983). *Generalized Linear Models.* 1st Edition. London: Chapman & Hall.

Preface

In the face of numerous blogs, Wikipedia entries, podcasts, YouTube videos, chatbots, etc., why a book on philosophies, puzzles and paradoxes? During most of our career, which started in the dark ages before the internet, we have actually been thinking about these things and the statistical issues underneath them, but never seriously enough to write anything about them. But now that we are at the end of our career, we realize that it would have been good for us to know and understand these issues earlier. Unfortunately, they are not the kind of topic found in standard textbooks, and it is difficult to put high confidence in generic blogs or Wikipedia entries. The many decades of experience we have gained as professional statisticians have helped us formulate and convey these topics with unique perspectives to new generations of statisticians.

In our effort to illuminate some poorly lit terrain of statistical inference, we realize that we're not offering easy-to-digest five-minute blog pieces. To paraphrase Mark Twain, we don't have time to write short pieces, so we write long ones instead. Our only defense is that most of the chapters here are based on many years of thinking, discussing and contemplating, so we hope the readers would indulge us by giving more than five minutes of their precious time. So, after all these meanderings, why a book? We believe that a book – perhaps even in its old-fashioned physical form – is still the best medium for reflective reading that these topics deserve.

Our primary interest is in the philosophical foundation of probability and statistics. It is a well-trodden area and some topics, such as the philosophical disagreements between the frequentist and Bayesian schools, have been debated for nearly a century. Serious students of statistics are likely to be perplexed by these disagreements, as they do not occur in other areas of mathematics or scientific disciplines such as physics, chemistry or biology. Unfortunately, although the passionate arguments have largely died down, it's actually not due to any consensus. Statisticians still argue about how to interpret the confidence interval and the role of P-value. Whatever truth that forms the foundation of our subject, it must be of the elusive kind that Oscar Wilde said to be rarely pure and never simple. Our recent realization of confidence as an extended likelihood, which leads to the concept of epistemic confidence, might contribute to a consensus based on the common ground between the epistemic non-Bayesianism and objective Bayesianism.

Many topics are presented via puzzles and paradoxes that highlight the nature of probability and statistical inference. The existence of a paradox is a warning that there is something incomplete, if not wrong, in our reasoning. Paradoxes have always been important in the development of mathematics; for instance, Russell's paradox about the set of all sets that are not members of themselves led to the development of set theory, logic and the foundation of mathematics. Similarly, in statistics, discussion and resolution of paradoxes may lead to a clarification of concepts and an enlargement of vocabularies that will strengthen the foundation of our subject.

Our book is an eclectic, some might even say idiosyncratic, collection of topics that reflect our interests. We believe that most chapters are readable by nonspecialists who have some knowledge of the elementary theory of probability and statistics, perhaps after one semester each. Several topics are indeed advanced, but the corresponding discussions rarely require deep knowledge or special technical competence. We have never used the book for any course, though we imagine that it could be useful for a discussion course.

Last but not least, we acknowledge the contributions of many colleagues and former students who, over the years, we have both challenged and, at times, tortured with these puzzles and paradoxes. Like jokes, puzzles and paradoxes are enjoyable for sharing. But unlike jokes, our understanding of puzzles and paradoxes improves with sharing. We thank Lars Rönnegård for careful reading of the draft and helpful suggestions, Zheng Ning and Jiangwei Sun for discussion and critical reading of several chapters on the paradoxes, and Jaehyuk Kwon and Dr. Hangbin Lee for editorial assistance. We have tried to present all facts, ideas and points of view accurately, but nobody is perfect, so we welcome corrections, comments or clarifications; these will be collated in `https://www.meb.ki.se/sites/yudpaw/book/`.

Yudi Pawitan and Youngjo Lee
Stockholm and Seoul

List of Abbreviations

CA countable additivity
CP conditionality principle
DF99 Donnelly and Friedman (1999)
FA finite additivity
HGLM hierarchical generalized linear model
GLM generalized linear model
iid independent and independently distributed
LLN law of large numbers
LP likelihood principle
MLE maximum likelihood estimate
RDF Ramsey and de Finetti
SLLN strong law of large numbers
SLP strong likelihood principle
SP sufficiency principle
VNM von Neumann and Morgenstern (1947)
WLP weak likelihood principle
WLLN weak law of large numbers

List of Puzzles and Paradoxes

Introduction and Summary

Even though we're writing a book for nonspecialists, we still realize that it would be impossible to satisfy the varying levels of all potential audiences. Advanced readers may find some materials too trivial, while others may find them too advanced. Some sections are starred as warning signs that they require advanced mathematical knowledge or use concepts that appear in later chapters. Nevertheless, at the cost of some repetitiveness, we have tried to make the chapters as self-contained as possible so that the reader can jump around following their interest. All the chapters in Part IV on the paradoxes, except perhaps Chapter 14 on the Allais and Ellsberg paradoxes, can be read independently from the rest of the book.

When we perform data analysis, there is a feeling that we're searching for the truth – the true model, the true parameter, etc. Mathematicians try to find and prove the true propositions; scientists in various fields work to discover the true laws of nature, the true causes of diseases, etc. Yet it has taken philosophers many centuries to understand the nature of truth and the methods to establish it. Chapters 1–4 are summaries of our attempt to absorb what (some) philosophers have thought and written about knowledge and truth.

Our primary interest is statistics and statistical reasoning. This interest leads immediately to probability, not just mathematically, but also philosophically. Although there is a consensus on mathematical probability around Kolmogorov's system of axioms, interestingly, even philosophers disagree on the meaning of probability. To some extent, but not fully, the Bayesian-frequentist divide reflects the philosophical disagreements. Philosophers spend their time thinking about these things more than statisticians do, so we find it worthwhile to read their literature closely. This is the basis of Chapters 5 and 6. We were rather surprised to discover, for example, that what we understand as the 'frequentist philosophy' actually does not follow from the 'frequency theory' of probability.

We have focused on orthodox concepts of uncertainty: probability, likelihood and confidence. Other concepts have been proposed, such as fuzzy logic, modal logic, belief functions, possibility and plausibility measures, etc. There is a large literature on these topics that is almost independent of probability, suggesting the enormous depth of the notion of uncertainty. Within probability-based inference, we cover in depth the basis of Bayesian inference by going into Savage's axioms in Chapter 7, historical records from Bayes and Laplace in Chapter 8 and the problem of prior selection in Chapter 9. Our own professional work and experience revolve around likelihood (Chapter 10), the central concept of statistical model and inference first explicitly recognized by Fisher in 1921.

During the 1930s, Fisher proposed fiducial probability, which was meant to give probability-based inference without having to specify prior distribution. Since it is not a proper probability, the fiducial idea led to many controversies, but it motivated the development of the confidence procedure (Chapter 11). In recent years, likelihood has been fundamentally extended to deal with random parameters (Chapter 12). The recognition that confidence is an extended likelihood leads to the epistemic confidence concept in Chapter 13. The rest of the book (Chapters 14–19) is dedicated to various puzzles and paradoxes. Short summaries of the chapters are as follows.

Chapter 1 covers philosophical theories of knowledge and truth, an area in which philosophers have traditionally been interested. They have the advantage of being professional truth seekers, so it seems natural that we should try to learn from what they have come up with. Two enduring schools of thought emerged since the Age of Enlightenment: the Continental rationalists, who emphasized reason, and the British empiricists, who emphasized the role of experience and observation. Hume's struggle with induction is particularly relevant for statisticians.

In Chapter 1 we see the recurring themes of necessary and contingent truths along with deductive and inductive reasoning to uncover them. In **Chapter 2** we go into more detail. Deduction, favoured by rationalists, uses pure reason to prove propositions based on known truths. It relies on assumed propositions for its foundation, but the validity of these premises can't always be observed. Induction, rooted in empiricism, involves evidence and support through experimentation. Inductive reasoning, though commonly used in daily thinking, remains tricky and doesn't necessarily improve with age. It's characterized by uncertainty due to incomplete information, leading to potential pitfalls. Clarifying the limits of both deductive and inductive logic is crucial for establishing reliable knowledge.

Throughout the centuries, mathematics has been seen as a source of objective truth achieved through deductive reasoning from self-evident axioms. **Chapter 3** discusses Hilbert's dream of automating all mathematical truths; it's a dream shattered by Gödel's incompleteness theorems. Together, the theorems highlight that, within a formal system, there are undecidable statements and true statements that cannot be proven, thus requiring reliance on other sources like intuition or empirical evidence. Uncertainty in deduction calls for additional methods to determine the truth. In geometry, different sets of axioms can lead to or represent different spatial models. Similarly, models in science and statistics, though not necessarily true, are valuable for understanding reality and making predictions.

The complex interplay between theory and observation in the scientific process is covered in **Chapter 4**. Some philosophers argue that theories or thoughts precede evidence, while others emphasize starting with observations to establish theories. Kuhn's model outlines stages of normal science, extraordinary research, paradigm adoption, and the establishment of new axioms. Paradigm shifts occur when dominant theories clash with new evidence or anomalies, leading to the adoption of new paradigms.

Chapter 5 charts the rise of probability as a model and a tool for dealing with uncertainty. Probability encompasses both deductive and inductive reasoning, offering a language to handle uncertainty. Its historical development involved diverse interpretations, from representing chances in games to epistemic degrees of belief. Kolmogorov's axioms laid the foundation for mathematical probability, distinguishing it from measure theory by introducing conditional probability and independence.

Although most statisticians follow Kolmogorov's mathematical probability, there is a wide range of philosophical theories and interpretations. **Chapter 6** covers: (i) classical, (ii) logical, (iii) subjective, (iv) frequency, (v) propensity and (vi) consensus theories of probability. The Bayesians adhere to the subjective theory, while the frequentists follow a blend of frequency and propensity theories. The consensus theory is a novel theory that gives the most compelling objective probability of single events.

Chapter 7 evaluates Savage's theory of subjective probability, which is based on axioms of rational preferences over acts and consequences. We describe in detail the seven axioms: weak ordering, sure-thing principle, state independence, consequence independence, non-triviality, small-event continuity and strong dominance. The meaning and implications of each axiom are discussed, as well as how they relate to other theories of decision-making. The theory highlights the power and limit of axiomatic systems, and how and whether they can be used to guide rational decisions under uncertainty.

Early writers like Pascal and Leibniz used probability to quantify uncertainty, often relying on simple reasoning similar to the principle of insufficient reason. In **Chapter 8** we discuss Bayes's 1763 *Essay Towards Solving a Problem in the Doctrine of Chances* and Laplace's 1774 *Memoir on the Probability of Causes by Events*, which laid the foundation for the so-called inverse probability method that dominated the 19th-century applied probability and statistics. Their idea solved Hume's struggle with induction, providing a rational approach to incorporating new information in probability estimation.

The Bayesian school is founded on subjective probability and the use of priors. The subjective probability can be justified by the coherency axiom, an attractive axiom even for non-Bayesians. However, the choice of a prior distribution remains problematic for most non-Bayesians. In **Chapter 9** we discuss how different choices of the prior distribution lead to different variants of Bayesianism, with 'objective Bayesians' using non-subjective and non-informative priors.

In **Chapter 10** we describe the concept of likelihood and explain how it differs from probability. Likelihood is central in statistical modelling and inference. Maximum likelihood estimates (MLEs) are widely used in practice, but in orthodox statistical inference, likelihood requires probability-based calibration. The key role of likelihood in statistical inference is based on its property of containing all the evidence about a parameter in the data, a result implied by the likelihood principle. However, a normative interpretation of the strong likelihood principle that demands equivalent likelihoods to yield identical inferences leads to controversy, especially among frequentists.

Significance testing and P-value concepts are covered in **Chapter 11**. Fiducial probability, introduced by Fisher via the P-value as a function of the parameter, resembles a standard probability. However, crucially, it lacks full additivity, which causes problems when dealing with many-to-one transformations. Although the fiducial idea has been abandoned in statistical practice, it motivated the widely accepted confidence interval procedure and the more recent confidence distribution concept.

In **Chapter 12** we describe extended likelihood and hierarchical likelihood as the unifying tools for dealing with complex models that involve fixed and random parameters. We explain how extended likelihood is different from classical likelihood. We show the difficulties that can arise when using extended likelihood, primarily due to the lack of invariance with respect to the transformation of the random parameters. Hierarchical likelihood is introduced as a special extended likelihood that avoids these problems and gives optimal estimators and predictors for fixed and random parameters. Confidence is shown to be an extended likelihood; this connection expands confidence's utility beyond the confidence-interval context.

Chapter 13 introduces the concept of epistemic confidence, an objective sense of uncertainty in unique events. For confidence intervals, it is the sense of confidence in a computed or observed confidence interval. Orthodox frequentist inference denies epistemic confidence, treating the confidence level as applying to the procedure rather than the interval. Epistemic confidence is essential for decision making and is naturally embraced by Bayesians. To make non-Bayesian confidence epistemic, we follow Fisher's proposal that a probability is epistemic if there is no relevant subset, but we use the confidence concept instead of probability. The epistemic confidence density and implied prior mirror the Bayesian posterior and prior, offering a non-Bayesian alternative to Bayesian inference. However, unlike Bayesian approaches, epistemic confidence doesn't require an explicit prior and isn't a probability.

In **Chapter 14** we discuss Allais's and Ellsberg's paradoxes, two well-known paradoxes associated with Savage's axioms (Chapter 7), particularly the sure-thing principle. The paradoxes highlight the descriptive/predictive vs normative roles of the axiomatic system. In the former role, the axioms should be able to predict human rational decisions, so a common violation of any of the axioms raises the question of whether the system is rich enough to capture real human behaviour. However, in its normative role, the axioms act as a guide to human decision making, where a violation would imply that it is the human decision that needs to be corrected in order to follow the axioms.

We use probability to quantify uncertainties in our reasoning for making judgements and decisions. Intriguingly, when used mathematically, probability-based thinking can sometimes produce starkly different conclusions compared to when used statistically. Our aim in **Chapter 15** is to discuss a well-known fallacy called the conjunction fallacy and a related inclusion fallacy. In the spirit of this book, they might be considered paradoxes, where seemingly valid probability-based reasoning leads to a contradiction viz-á-viz statistical reasoning. We will contrast two different modes of reasoning, one captured by probability and the other by likelihood, and suggest that our seemingly irrational behaviour is often due to decision making based on the likelihood. Thus, from a likelihood perspective, we are still acting rationally.

The Monty Hall puzzle has been solved and dissected in many ways, but always using probabilistic arguments, so it is considered a probability puzzle. In **Chapter 16** the puzzle is set up as an orthodox statistical problem involving an unknown parameter, a probability model and an observation. This means we can compute a likelihood function, and the decision to switch corresponds to choosing the maximum likelihood solution. We also describe an earlier version of the puzzle in terms of three prisoners: two to be executed and one released. Unlike goats and cars, these prisoners are sentient beings that can think about exchanging punishments. When two of them do that, however, we have a paradox where it is advantageous for both to exchange their punishments with each other.

Dealing with uncertainty is never easy. Although probability as the main tool for dealing with uncertainty is well developed, we still have to deal with many probability-related puzzles and paradoxes. In **Chapter 17** we describe the lottery paradox, which shows the logical problem of accepting uncertain statements based on high probability. In the cold suspect puzzle, the differences between the schools of inference – the frequentist, Bayesian and likelihood schools – lead to conflicting quantification of forensic evidence in the assessment of cold suspects from database searches.

In Hempel's paradox of the ravens, using seemingly logical reasoning, seeing a non-black non-raven – such as a red pencil – is considered supporting evidence that all ravens are black. This sounds non-sensical, but previous Bayesian analyses and Hempel himself accepted the paradoxical conclusions. **Chapter 18** describes likelihood-based analyses of various statistical models that show that the paradox feels paradoxical because there are natural models in which observing a red pencil tells us nothing about the colour of ravens.

We have explored the various interpretations of probability in Chapter 6. An intriguing paradox that arises in the context of probability for single events is the exchange paradox in **Chapter 19**. You are asked to choose and open one of two envelopes that contain some amount of money. Seemingly correct probability-based reasoning suggests that exchanging is always better, which is absurd, since you have just chosen one at random. The paradox highlights the role of likelihood, extended likelihood, confidence and subjective utility to deal with the uncertainty for single events.

Part I

How Do We Know What We Believe Is True?

1

Philosophical Theories of Knowledge and Truth

> Whoever undertakes to set himself up as a judge of Truth and Knowledge is shipwrecked by the laughter of the gods. – Edmund Burke (1729–1797), Anglo-Irish politician-philosopher.

'I'd believe it when I see it with my own eyes!' That sounds like a gold standard for verifying beliefs, but it's only for a hyperskeptic or an exasperated person. Certainly not the ultimate standard, as such a standard would limit our knowledge base severely. Let's consider a few questions,. some of which are intrinsically factual and directly verifiable, and some not: Did O.J. Simpson kill his ex-wife? Is Socrates mortal? Are protons immortal, or do they decay? Is there – or was there ever – life on Mars? Is there life somewhere else in the universe? Do the angles inside an arbitrary triangle always add up to two right angles? Does mask-wearing help reduce the spread of covid-19 infections? Does the mRNA vaccine work against the virus? Does ivermectin work? Even verifiable ones, such as the last three questions, require very careful and rigorous steps to answer, not verifiable just by our eyes. You may feel that you already 'know' the answers to some of the questions, but how do we know what we believe is true?

To answer the question, let's turn to the philosophers; they're the professional thinkers who spend their days contemplating the meaning of beauty, ethics, the purpose of life, or even the art of travel and the sources of value in economic commodities. We wish to find revelations from the ones that have thought deeply over several centuries about knowledge and truth. The wish turns out to be in the easier-said-than-done category. Perhaps it is not just the vigorous verbal gymnastics that makes us woozy. Unlike reductionist scientists, who are happy with incremental progress, philosophers have a tendency to build an all-encompassing theory of knowledge, but often with too few concrete examples.

But, of course, we couldn't give up and let you down, dear readers. After some perseverance, we identified recurring ideas that appeared over the centuries and made sure that we absorb the essential philosophical ideas critically. By way of apology, due to natural time constraints in this sojourn outside our comfort zone, we have focused unevenly on a few celebrated (and dead) philosophers whose names are synonymous with the study of knowledge and truth.

During the late 17th and early 18th centuries, Europe went through a special time known as the Age of Enlightenment. It was a period characterized by a focus on reason and the pursuit of knowledge and marked by significant advances in science, philosophy and politics. Many writers and philosophers from that era are still read today, similar to those who appeared during the Golden Age of Philosophy in Greece during the fifth and fourth centuries BC, such as Socrates, Plato and Aristotle. Perhaps it's not surprising that probability theory also emerged during this period, and many major figures of the Enlightenment era contributed to probability theory.

The Enlightenment philosophers fall roughly into two camps: continental rationalists and British empiricists. Rationalists – such as Descartes, Leibniz, Kant or Hegel – were either mathematicians themselves or highly influenced or even enthralled by mathematics and science. They emphasized the role of reason and logic to arrive at knowledge and truth. On the other hand, empiricists, such as Locke, Berkeley or Hume, were less influenced by mathematics and placed more emphasis on experience as the source of knowledge.

Those definitions are, of course, rough and simplistic. Rationalists do not claim that knowledge is independent of experience, but rather that reasons provide a framework that structures the experience, or provide concepts that give more information than experience alone. And perhaps most importantly, reason is the final arbiter of truth. On the other hand, empiricists would consider two kinds of experience, sensory and reflective, where the latter is not far from rationalist reason. So in reality, in any single person, there is no clean division between the two traditions: Descartes had empiricist ideas, whereas, for issues such as ethics, Locke weighed reason and experience equally. Hume also recognized the power of deductive logic in establishing 'Relations of Ideas' such as those found in mathematics.

1.1 The Rationalists

René Descartes (1596–1650), considered the father of modern western philosophy, is best known for his *Meditations on First Philosophy* (1641), where he sought to provide a foundation for knowledge and sure belief. Gottfried Leibniz (1646–1716) perceptively identified two kinds of truth: necessary and contingent.[1] A necessary truth is one whose negation leads to a contradiction. Mathematics is the obvious place for necessary truths, which can be demonstrated by reason alone. On the other hand, a contingent truth is not necessarily true, but is nonetheless true. Negation of a contingent truth does not lead to a contradiction. True empirical facts are always contingent. For instance, at present the sky is blue, but of course it doesn't have to be: it could be gray and rainy, especially as we're writing this very sentence in Ireland.

Truth is established based on the great principle of rationalism: the principle of sufficient reason, which states that there is a reason or cause for everything, including the existence of objects, the occurrence of events, and the truth or untruth of propositions. It's actually an ancient principle already known to the Greeks. Necessary truths must be true in order to avoid contradiction. Contingent truths must also have some underlying cause or reason. If your car breaks down – an annoying contingent truth – you'd be sure there must be a reason, which a mechanic should be able to find out and fix. We'd have a similar reaction to bigger problems, such as the ozone hole over Antarctica or the mass death of coral reefs in the world's oceans. The rationalist goal is to uncover the cause or explanation of any phenomenon through reason and inquiry. We inherit this optimistic view of science to this day.

If everything has a reason, then there must be a chain of reasons going all the way to the ultimate reason. Being a devout Christian, for Descartes the ultimate reason is God, whose existence he proved deductively (Section 2.2). For Leibniz, it is the 'monad,' the fundamental building block

[1]During the 20th century, we saw a formalization in modal logic in terms of necessary and possible truths with its colourful possible world semantics. Thus, a necessary truth is true in all possible worlds; a possible truth is true in at least one possible world; a contingent truth is true in some but not all possible worlds. Gödel's theorem on God's existence in Section 2.2 is expressed in modal logic.

of the universe, all reality and knowledge. It is not clear what a monad is; philosophers did not even seem to agree with each other about it.

In deductive logic, all propositions have to rely on previously established true propositions, so to break the infinite regress, we have to start with propositions that do not require a proof. These are the axioms, whose nature of truth remained somewhat mysterious and contentious until the 20th century. The early rationalists simply assumed that, by the power of reason, we could recognize the axioms as self-evidently true. In modern mathematics, this position is no longer common. Axioms are now considered true only by definition or by construction.

Immanuel Kant (1724–1804) is widely regarded as one of the most influential figures in western philosophy. We shall give him a lot of space befitting his importance. Writing about 150 years after Descartes's *Meditations*, in his *Critique of Pure Reason* (1781; second edition 1787) he lamented the state of philosophy ('metaphysics') which, he said, was ruled by 'the antiquated and rotten constitution of dogmatism.' To bring things into order, he introduced the concepts of (i) analytic vs synthetic knowledge, and (ii) a priori vs a posteriori knowledge. (Kant used the word 'judgement' or 'cognition,' not exactly the same as knowledge, but for uniformity we will use 'knowledge.')

A priori knowledge is independent of any experience whatsoever, as opposed to a posteriori knowledge, which must be established by sensory experience. Analytic knowledge is tautological; it's true by construction or by definition, not by checking against reality. For instance, the sentence 'All bachelors are unmarried' is analytic. It is certain, but the certainty comes with a price: It does not contain any factual information. On the other hand, synthetic knowledge is non-tautological: It's built from a combination of two or more unrelated concepts in a new way, so it's not necessarily true. For instance, 'All bachelors are happy' is synthetic, since happiness is not built in the concept of bachelorhood, and its truth needs to be checked against reality. Kant argued strongly that the statement '7+5=12' is synthetic, supposedly because the idea of '12' is not contained in '7' or '5.' It's a highly questionable view, since surely no one can find any factual evidence that can negate the truth of 7+5=12. Indeed, the 20th-century logical positivists considered it analytic, as it is a consequence of Peano's axioms of arithmetic together with the definition of the '+' operation (Hempel, 1945a).

It may help us to understand Kant's theory more easily if we put his analytic-synthetic concepts in parallel to Leibniz's necessary-contingent truths. We might concede a slight difference between analytical and necessary truths: Axioms are analytically true, as they are true by definition. Whereas, given axioms, theorems are necessarily true, since their negation implies a contradiction. But we could also argue that theorems are true by construction, hence analytic. In any case, the synthetic and the contingent truths look and smell the same.

Kant's great philosophical question was whether synthetic a priori knowledge is possible. To be a priori, knowledge has to be necessarily and universally true, and innate in the human mind, that is, not abstracted from experience. But, to be synthetic, it has to combine unrelated concepts, so it's not necessarily true. So 'synthetic a priori knowledge' actually sounds contradictory, but he argued that we do have synthetic a priori knowledge. In mathematics, such knowledge is contained in the principal rules of logic and, at the time, Euclid's axioms in number theory and geometry. To some extent, he continued, such knowledge is also found in the basic laws of physics, e.g., Newton's laws or the law of conservation of mass. In effect, Kant elevated the lame 'self-evident' qualification of axioms to a philosophically sophisticated 'synthetic a priori' label.

His ultimate goal was to reconstruct metaphysics using only synthetic a priori knowledge, with the expectation that it would be as pure and universal as mathematics. He identified such axiomatic knowledge in our understanding of space and time, and our grasp of cause and effect, which are crucial for organizing and making sense of our sensory experiences. Kant's ideas about space and time have been understood as innate forms of perception, known as outer and inner intuition.

As these innate abilities are not learned from experience, they are presumably present at birth. Due to the lack of clear examples, it is uncertain whether his metaphysical concept of synthetic a priori knowledge falls under the category of what is commonly called instinct. For spiders, the ability to weave a web must be synthetic a priori, since it's completely innate and independent of experience. It's coded in their genes. Similarly, human linguistic skill, especially grammar, is an instinct, an innate ability encoded in the genes that shape our brain. Exposure alone does not explain children's depth of understanding and mastery of the grammar of their native language. Our intuitive grasp of cause and effect also appears to be encoded in the brain: Damage to the frontal cortex can hamper or even erase our causal reasoning.

But all these instincts are products of evolution, a long-term adaptation based on experience living in a *specific* natural environment. During critical periods in childhood, our plastic brain needs environmental input in order to develop properly. For example, a deaf child will grow up mute due to lack of sound input. In ground-breaking studies that eventually led to the Nobel Prize, David Hubel and Torsten Wiesel in the 1960s showed that, during its third to eighth week, a kitten must receive visual stimulation to avoid blindness. Hence, our mental faculty is not a priori in the Kantian sense, which requires absolute independence from experience. The main problem with Kant's theory is that it depends so much on mental cognition, but the knowledge of neuroscience in the 18th century was virtually nonexistent.

Yet, Kant's theory of knowledge survives to the present day, with associated books and papers still being published, imposing itself on those genuinely interested in the philosophy of knowledge. He was an energetic obscurer of language: Reading him, if you have to, should only be done when your brain is in its freshest state. He loved befuddling words like 'apodeictic' or 'noumena.' The word 'transcendental' appears more than 500 times in the *Critique*, often in unexpected combinations with no obvious meaning. Here is the reason:

> I apply the term transcendental to all knowledge which is not so much occupied with objects as with the mode of our cognition of these objects, so far as this mode of cognition is possible a priori.

Thus, 'transcendental aesthetic' refers to our knowledge of space and time, while 'transcendental analytic' refers to our knowledge of cause and effect. Subsequent philosophers disagree with what he meant in his basic pronouncements, for instance, on the nature of space. His fundamental views of the axioms in geometry and our sense of space and time proved to be incorrect, as clearly seen after the establishment of non-Euclidean geometry and Einstein's theories of relativity. Logical positivists in the 20th century do not consider mathematical axioms as synthetic, but analytic; see more below. Any physicist today will have a completely different intuition about space and time compared to 18th-century physicists, so our mind does not have an a priori notion for space and time.

With the 'transcendental idealism' (or 'ideality' in some translations), Kant claimed that our mind actively constructs our experience and imposes its own concepts and categories on sensory input from the external world. In other words, all empirical objects are mind-dependent: What we experience is not the 'world in itself'; we cannot know 'things in themselves,' but only our mind's representation of it. To some extent a similar 'idealism' also appears in N.R. Hanson's and T. Kuhn's contention in the 20th century that our observations are 'theory-laden.' That is, our effort to understand the world is shaped by previous theories or models created in our mind to make sense of it.

We can view Kant's idealism from two opposing perspectives. First, from the biological perspective, it is actually very close to how modern neuroscience views our understanding of the world as a product of our brain's processing of external input. It is well known that our brain has a 'mind' of its own, which, disturbingly, can be independent of reality. One of the most dramatic examples is the phenomenon of the phantom limb (Doidge, 2008), where an amputee still feels the presence of the amputated limb and even feels pain on it. But, from the second perspective, at least in physics, to say that we cannot know 'things in themselves' sounds anti-scientific, since the purpose of science is to explain nature objectively, i.e., independently of our mind. The scientific view is more like this: There is no limit – or, we do not know the limit – in our understanding of nature. The relativity and quantum theories have revealed strange properties of nature that are far beyond a priori human cognition. Physicists would say, that's the way nature is.

In summary, the primary lesson we get from the rationalists is a special respect for the 'mind,' with its powerful reasoning, truth-preserving deductive logic and independent existence from the body. Descartes's model of mind-body duality survived well into the 20th century. In this model, the brain is an inert physical machine run by the lively mind or the soul – the ghost in the machine. Alas, the ghost was never found. In recent times, neuroscience has broken this Cartesian duality and shown that the mind is just a result of brain anatomy and processes. This implies that the mind has no special access to the truth a priori.

1.2 The Empiricists

With his emphasis on experience and observation as a source of knowledge, Aristotle is considered the father of empiricism. His 10-volume *History of Animals* alone contained detailed observations and descriptions of over 500 species of animals, covering their anatomy, reproduction, behaviour, habitats, etc. In addition, he wrote four other volumes on animals and at least eight volumes on physics.

However, Aristotle's credibility as an empiricist is undermined by glaring unforced errors that could be checked easily, such as men have more teeth than women,[2] and by fanciful theories such as this from *The Generation of Animals*:

> Again, more males are born if copulation takes place when north than when south winds are blowing. For in the latter case the animals produce more secretion, and too much secretion is harder to concoct; hence the semen of the males is more liquid, and so is the discharge of the catamenia [menstruation].

Bertrand Russell (2016, p. 7) remarked sarcastically that 'although [Aristotle] was twice married, it never occurred to him to verify this statement by examining his wives' mouths.' Was Russell just cherry picking? A classicist, Peter Gainsford, checked 21 statements Aristotle made about teeth: he was 83% correct.[3] That's an excellent mark for a student in an exam, but perhaps not good enough for a reference text by a great philosopher. For instance, which of the following statements from *The History of Animals* would you trust: (i) Elephants have two tusks: large and bent upward in the case of males, small and bent downward in the case of females. (ii) Elephants have teeth at birth?

Did Aristotle not care about the truth? In the passage on male vs female births in *The Generation of Animals*, he actually wrote: 'Observed facts confirm what we have said.' So, he was aware of the 'facts' and he was not making things up. The problem was obviously in the poor standard of proof of empirical statements. It's an interesting historical question

[2]Full quote from *The History of Animals*: 'Males have more teeth than females, in the cases of humans, sheep, goats, and pigs. In other species an observation has not yet been made.'

[3]`http://kiwihellenist.blogspot.com/2017/09/aristotles-errors.html`.

in itself, but in any case it highlights that the meticulous fact checking in the culture of modern science is actually an achievement we cannot take for granted.

In *Metaphysics*, Aristotle discussed an early version of the principle of sufficient reason, also called the principle of causality: Things exist or happen for a reason, which can be discovered by observation. This was a move away from Plato's idealism, which considers the ultimate reality to be an eternal non-material realm – the so-called Platonic world – of perfect Ideas and Forms. For an extreme empiricist, all knowledge comes from experience. But real empiricists, such as Aristotle or Hume do consider mathematics, particularly geometry, as a product of thinking. Aristotle also believed in ultimate reasons that are not verifiable by experience or observation.

Aristotle had an interesting view on knowledge that is based on convincing others rather than oneself. In his *Rhetoric* he wrote that a speaker has three ways to persuade his audience: by the power of his personal character, by appealing to their emotion, or by providing arguments. Of the arguments, he provided a demonstrative/deductive logical argument or an inductive argument based on examples or analogy. As for the demonstrative arguments, they must be based on some organizing principles – the axioms. He spent a lot of time explaining how to arrive at the principles, where he emphasized the role of experience and induction.

In the *Enquiry Concerning Human Understanding*, a remarkable 121-page book first published in 1748, Hume (1711–1776) wrote (p. 16) what's clearly a reaction to the continental rationalists and emphasized the importance of sensory experience:

> All ideas, especially abstract ones, are naturally faint and obscure: the mind has but a slender hold of them: they are apt to be confounded with other resembling ideas.
>
> On the contrary, all impressions, that is, all sensations, either outward or inward, are strong and vivid: the limits between them are more exactly determined. When we entertain, therefore, any suspicion that a philosophical term is employed without any meaning or idea (as is but too frequent), we need but enquire, from what impression is that supposed idea derived? And if it be impossible to assign any, this will serve to confirm our suspicion.

For scientists, it is easy to side with Hume. It's ironic that Kant credited Hume for waking him up from his 'dogmatic slumber,' as he became a prolific inventor of obscure philosophical terms.

Hume (p. 18) categorized all subjects of human reasoning into two different types: (i) Relations of Ideas, which include geometry, logic and arithmetic, where every affirmation is either intuitively or demonstratively certain. The truth here is 'discoverable by the mere operation of thought, without dependence on what is anywhere existent in the universe.' (ii) Matters of Fact, a completely different subject from the first, since their negation never implies a contradiction, so they are not necessary truth. This division has close parallels to Leibniz's necessary and contingent truths.

Hume clearly recognized the need for special reasoning to deal with facts, which we now see as inductive reasoning. Though he himself failed to solve it, it is instructive to follow closely Hume's thoughts and open struggle. The principle of sufficient reason would only state that there must be a reason underlying the matters of fact, but it does not provide any method of reasoning. Hume realized that a matter of fact must be established by another fact; all reasoning concerning them relies on the causal relationship; and the causal knowledge comes not from thinking a priori, but from experience.

To appreciate Hume's difficulties, it's instructive to recognize the different layers of 'Facts' in terms of verifiability. There are simple facts, the kind reported in newspapers, that are easily verifiable and those that are not. The correlation between smoking and cancer is a fact that is easy to verify, but whether 'smoking causes cancer' is not so easy to verify. Also easy to verify is a claim that when John was very sick with covid-19, he improved after taking ivermectin, but it'd be impossible to verify that it was ivermectin that cured him. It could just be due to the passage of time. It's more feasible, though still a nontrivial effort via a rigorous clinical trial, to verify whether ivermectin has any curative benefit in a group of people. Seeing a flock of black ravens is a simple fact, but the claim that 'all ravens are black' is no longer a verifiable fact. Indeed, these claims should actually be considered ideas, more specifically 'contingent ideas,' whose truths are only contingent.

In general, facts that are unverified, difficult or virtually impossible to verify directly, such as the composition of the Earth's core or the properties of black holes or the existence of dark matter, resemble theories – hence ideas – rather than mere 'Facts.' All statistical hypotheses and scientific theories, and their unverified predictions, are contingent ideas; calling them 'Facts' just sounds wrong, even though we use facts to establish them or in the future some of them might be considered facts. So Hume's hard separation of Ideas and Facts is too simplistic: Conceptually, in fact, the trickiest reasoning problem is found in the Relations between Facts and Ideas.

Mathematics and deductive logic are used to establish necessary truths based on the Relations of Ideas. But what kind of reasoning can be used to establish the truth of a contingent idea? Once we leave deduction, we're on an uncertain ground. Unlike deduction, induction is not truth-preserving: Correct premises/facts do not necessarily lead to correct conclusions. As these conclusions are often in the form of theories that aim to explain the facts, we may say that theories are underdetermined by facts. Even the most bizarre conspiracy theory is based on facts, albeit selective. Epidemiologists have 74 names for different biases lurking when we try to draw conclusions from data (Delgado-Rodriguez and Llorca, 2004). Not all of these biases are distinct, but they nonetheless highlight a high number of potential pitfalls that we can encounter when analysing observational data. To make matters worse, in some cases, induction is discontinuous: adding a new fact could lead to a dramatically different conclusion; just think of the crime stories. As a consequence, overall, there is no inductive method that can guarantee the truth of its conclusions.

Hume spent a large portion of his book on ideas or propositions in which we gain confidence from repeated occurrences of simple verifiable facts. It's a pleasant monotone property of a certain type of induction problems that actually makes them amenable to a proper deductive analysis.

Let's consider a modern version of his example. You sometimes feel nauseous, bloated and constipated. You guess it must be food related, but since you're an omnivore and have never heard of coeliac disease, you can't figure out the cause. After some trials and errors, you're suspicious of bread and now start paying a close attention. The next time you eat it, you notice the symptoms. Aha, that's the culprit! Whenever you eat bread again, you are going to expect the same experience, though that's not necessarily true because the symptoms maybe due to some types of bread only. Your confidence in the cause of your problem will grow the more times it occurs. Here comes his key question: If causal knowledge comes from experience, what kind of reasoning is used to establish the knowledge? Hume (p. 25) seemed genuinely puzzled:

> But if you insist that the inference is made by a chain of reasoning, I desire you to produce that reasoning. The connexion between these [cause-effect] propositions is not intuitive. There is required a medium, which may enable the mind to draw such an inference, if indeed it be drawn by reasoning and argument. What that medium is, I must confess, passes my comprehension; and it is incumbent on those to produce it, who assert that it really exists, and is the origin of all our conclusions concerning matter of fact.

In other words, he was asking an inductive inference question, which was not yet available until Bayes and Laplace came to the scene later in the century (Chapter 8). Hume (pp. 26–28) could only argue the following:

> In reality, all arguments from experience are founded on the similarity which we discover among natural objects, and ...
>
> From causes which appear similar we expect similar effects. This is the sum of all our experimental conclusions.
>
> Now it seems evident that, if this conclusion were formed by reason, it would be as perfect at first, and upon one instance, as after ever so long a course of experience. But the case is far otherwise.
>
> It is only after *a long course of uniform experiments in any kind* [our emphasis], that we attain a firm reliance and security with regard to a particular event.

Hume was often said to be skeptical of induction, but he actually stated explicitly that he was not a skeptic:

> Now where is that process of reasoning which, from one instance, draws a conclusion, so different from that which it infers from a hundred instances that are nowise [sic] different from that single one?
>
> This question I propose as much for the sake of information, as with an intention of raising difficulties. I cannot find, I cannot imagine any such reasoning. But I keep my mind still open to instruction, if any one will vouchsafe to bestow it on me.
>
> As an agent, I am quite satisfied in the point; but as a philosopher, who has some share of curiosity, I will not say scepticism, I want to learn the foundation of this inference.

Let's call this Hume's challenge. However, due to a lack of inductive inference methodology in his time, Hume (p. 31) could only come up with a simplistic and wholly unsatisfactory explanation to the induction problem by appealing to the 'principle of Custom or Habit':

> All inferences from experience, therefore, are effects of custom, not of reasoning.

The growth of statistics and scientific methods can be seen as a systematic effort to provide rigorous methodologies to answer Hume's challenge.

1.3 The Positivists and the Verification Principle

During the 19th century, the positivists – such as Auguste Comte and John Stuart Mill – continued to focus on empirical evidence, but with more emphasis on scientific methods, particularly in the natural and social sciences, as the basis for understanding and explanation. In the 20th century, positivism was further developed by the philosophers of the Vienna Circle into logical positivism or logical empiricism. The main point of view is that philosophy must use the same empirical methodology as other sciences. In other words, there is no methodology that is unique to philosophy. Verification is taken as a central principle, to the point that what cannot be verified empirically is meaningless. Only mathematical and linguistic questions can be answered without reference to empirical support. Traditional metaphysical questions are out as far as true knowledge is concerned. Again, here we see the clear division between necessary and contingent truths.

Although positivists are all empiricists, compared to early empiricists such as Hume, the logical positivists pay more attention to mathematics and logic and place more emphasis on linguistics. The latter is due to Wittgenstein's influence. Non-Euclidian geometry, the axiomatization of mathematics, Gödel's incompleteness theorems (1931), etc., have contributed to the new perspectives. Euclid's axioms are no longer seen as self-evidently true a priori, but simply as assumptions for an abstract spatial model. Pure geometry, arithmetic, etc. are seen as a collection of analytic statements, which are only true *by definition.* So, strictly speaking, we can't say that Euclid was right or wrong, but only whether his geometry is consistent or not; for further discussion, see, e.g., Hempel (1945a). When geometry is applied to the real world, it becomes applied mathematics, part of the empirical sciences, where assumptions and models have to be checked against reality. Hence, assumptions are no longer analytic statements, but contingent or synthetic.

The views of logical positivists are recognizably scientific, but some philosophers such as Karl Popper or Bertrand Russell criticized them as being too narrow. Who will deal with questions of ethics, aesthetics or the meaning of life? The verification principle is also often interpreted loosely and hence unspecific. What exactly does it mean? In the certain world of mathematics, an analytic statement is 'verified' when it is demonstrated in a proof, which must consist of a finite number of steps only. Completeness and finiteness of the proof are essential, not optional, requirements. So, ideally, we should also have the same requirements for contingent truth. But, in reality, as we have described before, there are many levels of verification.

Commonly solvable things in life often satisfy the requirements, though not always. If you're asked to verify your identity by the authorities, you can do so completely in a finite number of steps, e.g., you can show your passport or utility bill. If there is something wrong with your car, a good mechanic can usually find the cause in a finite time and verify it by fixing it. Political opponents during an election campaign would verify each other's claims, and it's in their interest to do it completely and immediately.

However, if something is wrong with your body – a much more complex system – a doctor may take a long time to get a proper diagnosis; in some cases, there may not be a definitive diagnosis. And, the putative diagnosis may not be verified if the supposed treatment does not work. Or you may get a diagnosis with the adjective 'idiopathic,' such as idiopathic scoliosis, which actually means 'unknown cause.'

In court, a place where truth is supposed to be paramount, verification is also not a simple matter. Did O.J. Simpson kill his ex-wife? Whose evidence or arguments do you trust? The defence's or the prosecutor's? And what verdict is correct? In the criminal court, Simpson was declared not guilty. But the subsequent civil court ruled against him. 'Verification' is judged with the qualitative threshold of 'beyond reasonable doubt' in the criminal court, but only with the 'balance of probability' in the civil court. In both courts, various technicalities apply regarding what evidence can be included, etc.

Is science better at verification? Let's take an example: How do we verify that the core of the Earth consists of iron? We can never verify it directly. Is it enough to verify indirectly, for instance, by inferring it from the Earth's magnetic field? Much more is now known about the Earth's core: its composition, temperature, pressure, etc., all necessarily derived from indirect evidence. Indirect verification in science is indeed common, so while scientists often consider their theories as 'verified,' the theories are only considered provisionally true.

How would you verify the continental drift theory, as proposed in 1912 by Alfred Wagener, that the continents started as a single land mass and later moved apart? If you treat the continents as jigsaw puzzle pieces, some pieces, such as South America and West Africa, or East Africa and India, would fit pretty well. But, although you can't imagine how such a fit could occur by accident, it was only considered as circumstantial evidence, not sufficient to verify such a remarkable claim. Wagener himself proposed various indirect evidence, but for 50 years geologists rejected the theory because there was no good causal/mechanistic explanation. Keith Runcorn's research on paleomagnetism and polar wandering in the 1950s revived the continental drift theory. The convincing 'verification' came only when Harry Hess showed evidence of seafloor spreading around 1960.

Even if a theory has only a finite number of consequences, not all can or will be checked for the theory to be considered verified. A rich theory, such as the general theory of relativity, generates many testable predictions over a long period of time. To validate his general theory of relativity, Einstein made three predictions: (i) the anomaly of the perihelion movement of mercury; (ii) the deflection of light during solar eclipse; (iii) the gravitational redshift. These were verified in 1915, 1919 and 1954, respectively. At what point is the theory considered 'verified'? Physicists paid attention to Einstein's theory after the verification of the first prediction, and Einstein became a global celebrity-scientist after the second verification. The gravitational redshift has continued to be verified until recently, involving better precision or data from more and more galaxies. Other predictions of the theory– such as the expanding universe, the black holes and the big bang – came later, each time requiring more and more time to verify. So, in general, 'complete verification' is not required in science.

Even a simple statement such as 'All humans are mortal' is not strictly verifiable, since it will require the verifier to wait until the end of the human race. This violates the completeness and finiteness requirements. We cannot say 'All humans so far are mortal.' Who knows, among those alive today, some may be immortal when scientists discover the secret of immortality. Do we have to qualify the statement to something like 'All humans born before the year 1800 are mortal'? This is of course silly, since we might as well say 'All humans that have died are mortal,' so its truth is no longer contingent, but necessary and contains no information.

How do we verify that mRNA vaccines work against covid-19? First of all, it is not required that the statement is true for all patients, for all variants of the virus and at all times. As long as it is better than placebo by a certain margin, for a relevant variant of the virus at the present time, the statement is considered verified by the authorities and, we hope, by the public. Even with these lower standards, the clinical trials required to verify the statement convincingly were a multi-billion dollar effort.

In summary, like Hume with his 'Matters of Fact,' logical positivists overestimate our ability to verify, so tying meaning and truth to verification becomes problematic. Instead of 'verified,' Carnap introduced the notion 'confirmed,' where confirmation does not require a full verification (Section 8.6). At the other extreme, Popper suggested that, to be meaningful, a scientific hypothesis does not have to be 'verified' or even 'verifiable'; it only needs to be 'falsifiable.' This view is closer to the scientists' real modus operandi with their provisional theories.

2

Deduction and Induction

To sum up our philosophical explorations in the previous chapter, we can point to the recurring theme of necessary and contingent truths, along with the deductive and inductive reasoning that go with them. We shall now go a little deeper into these modes of reasoning. Deduction is the primary tool of the rationalist. In this mode of reasoning, one applies pure reason to prove correct propositions and to disprove incorrect ones, starting with central propositions presumed to be true. Euclid's geometry is seen as the epitome of a deductive axiomatic system. Induction is also known as analogy, examples or generalization. It is based on the empiricist tradition developed by Locke and Hume, and later by statisticians as well as natural and social scientists. With induction, one does not 'prove' an objective truth, but only provides evidence or support through experimentation and observations. Philosophers seem to recognize another form of reasoning called 'abduction,' but as we explain in Section 8.6, it is just a form of induction.

Deductive reasoning is highly attractive as it leads to certain conclusions. This might explain why we're drawn to generalizations, which can then function as basic premises in the deductive process. It might be argued that in a deductive process, no new information is produced, only seemingly tautological consequences of old information. Whereas in induction, the explanatory hypothesis can be a new creative inference suggested by the data, so the process generates new knowledge. But let's think more carefully: Theorems in geometry *are* consequences of the axioms, but they constitute vast new knowledge. We're all interested in our bank statements, especially in the current balance, even though it is just computed from other numbers following the deductive rules of arithmetic. So, deduction does produce interesting new information. What's true is that deduction does not produce new *factual* knowledge; this is fine, as not all knowledge is factual. Results from thinking alone can produce astonishingly rich knowledge beyond what's suggested by observations. In physics, starting with fundamental laws as axioms, one can derive an amazing array of novel theories and predictions, such as the movement of planets, $E = mc^2$, gravitational waves, black holes, etc.

Among the questions at the start of Chapter 1, there is one simple one: Is Socrates mortal? If this is Socrates the philosopher, then we know that he is dead, so we can answer in the affirmative. But you may have a neighbour named Socrates who is still very much alive. Yet, without waiting for the news of his death, we also *know* he is mortal. We have intuitively used a basic deductive reasoning known as Aristotle's syllogism, a logical argument where a conclusion is drawn based on two assumed propositions: (i) a major premise, taken from universal laws, scientific theories, axioms in mathematics, basic propositions in philosophy, creeds in religions, etc.; (ii) a minor premise, stating a specific instance, evidence, data or observation. Thus, in this case we intuitively use a general proposition as the major premise

G: All humans are mortal,

and the minor premise is that Socrates is a human. To obtain a specific proposition as the conclusion,

E: Socrates is mortal.

As long as we accept the two premises, we must accept the mortality of Socrates; there is no need for direct evidence.

In deductive reasoning, specific propositions follow from the general proposition. So, if we want to establish a specific proposition E, we look for a general proposition G whose truth can be easily accepted. On the other hand, the specificity of E means that it can often function as a testable prediction of the general proposition. Logically, if E is false, then G is falsified. If there existed an immortal human, then G cannot be true. (Here we see the formality of a deductive system, as it depends on or assumes unambiguous definitions. In reality, our idea of mortality may vary depending on how we view our existential self: If we could preserve our cell/genome indefinitely so that we can be cloned at any time in the future, would we be considered immortal? This has already been achieved with some animals, such as dogs or even camels. What if, additionally, we could also find a way to preserve our core personality, memories and consciousness?)

Although deductive reasoning is natural in mathematics, it also forms a large part of physics, with known or accepted laws taking the role of general propositions. Highly accurate knowledge of the planetary mass, positions, motion, etc. can be derived from Newton's laws. Let's consider the question: Is the proton immortal? Matter consists of protons, neutrons and electrons. Theoretically, the laws of conservation would imply that electrons are immortal, but there are no laws against the decay of protons and neutrons. Experimental evidence has indeed shown that free neutrons outside the nucleus decay within minutes. So far, there is no evidence to show the decay of protons. Which fact would violate the laws of physics: the immortality or the mortality of protons? Mathematically, the so-called Grand Unified Theories actually predict that protons must decay. It is, in fact, the strongest testable prediction of the theories.

In deductive reasoning, as long as G is assumed true, E must be true. But how do we know that G is true? Observing that E is true does not establish the truth of G. General propositions such as G cannot be entirely observed because we can't wait to see the death of all human beings. What we can observe is evidence (specific event) of G. For example, we see that a particular person, named Socrates, died at some point and we can confirm his death officially. This is the problem we face in deduction: the construction of the basic propositions and the fact that the very foundation of our deduction – the truthfulness of basic propositions – can be in doubt. Early rationalist philosophers simply claimed that there are self-evident true general propositions.

For 'Matters of Fact,' Hume argued against the existence of innate, a priori true propositions, positing that all human knowledge is founded solely on experience that serves as evidence, not from theoretical deductions. Furthermore, inductive reasoning and the belief in causality cannot be rationally justified via deduction. Instead, our trust in causality and induction results from our customs and mental habits and is only attributable to our experience of the 'constant conjunction' of events. This is because we can never actually perceive the causality of events. We can only perceive that two events are conjoined. Hume's skepticism had a great impact on the development of justifiable epistemic knowledge.

Unlike Hume's skepticism, which challenged the possibility of attaining certain knowledge, fallibilism does not reject the existence of knowledge. Rather, fallibilists acknowledge that our current understanding of the world may be incomplete or mistaken, and that even our most firmly held beliefs are subject to revision in light of new evidence or arguments. No belief can have a justification that guarantees its truth. There is no method to establish the truth of knowledge claims once and for all. Moreover, the notion of justification itself is a misconception, as it assumes that knowledge can be deemed genuine after being justified by a particular source or criterion. The 19th-century American philosopher Charles Peirce coined the name fallibilism in response to what he considered dogmatism during his time. We could see a strong trace of fallibilism in Karl Popper's view of scientific truths as provisional. Thomas Kuhn (1962) is also a fallibilist in his argument that scientific knowledge is not simply a collection of facts but is shaped by paradigmatic frameworks and the paradigms are always open to revision in response to an accumulation of anomalies.

In the early 20th century, many books on the foundations of statistics, ethics, justice, fairness, etc., were published. All of these books study foundational principles for each of their subjects and enable deductive logic to justify necessary propositions in their respective fields. We may call a change of foundational principle a paradigm shift in the subject area. However, how can we justify the truthfulness of the claimed general propositions? In religion, they are called creeds or matters of belief; if one does not believe them, one is not a believer of the religion. What about basic propositions in mathematics and sciences? Are they matters of belief or inductively justifiable knowledge?

To illustrate the problem of induction, Bertrand Russell, the famed British philosopher-logician-mathematician, gave the following example. There lived a very intelligent chicken, who knew it was to be fed every morning. So, it planned to announce its punctual breakfast to the public. Sadly, it never got to make the announcement because on the very day of the planned announcement, it was slaughtered and ended up as its owner's breakfast. As seen in this example, to draw any causal inferences from past experience alone, it is necessary to presuppose that the future will resemble the past, but this presupposition itself cannot be grounded in prior experience. This chicken example shows that observations alone, regardless of how many there are, do not guarantee our perceived future.

The difficulty surrounding the induction problem has been recognized since the Greek and Roman periods. A Pyrrhonian skeptic, philosopher and physician, Sextus Empiricus, questioned the validity of inductive reasoning:

> A universal rule could not be established from an incomplete set of particular instances. Specifically, to establish a universal rule from particular instances by means of induction, either all or some of the particulars can be reviewed. If only some of the instances are reviewed, the induction may not be definitive, as some of the instances omitted from the induction may contravene the universal fact. However, reviewing all the instances is impossible if the instances are infinite and indefinite.

Detecting causal effects by means of induction can be even more difficult. The 20th-century mathematician turned meteorologist Edward Lorenz found that, for a complex non-linear *deterministic* system, a micro perturbation in the input may cause a macro effect in the output. Metaphorically, this is called the butterfly effect: a butterfly flapping its wings in New Delhi may cause a hurricane in New York a couple of years later. If so, how can we figure out the causes of the hurricane with any precision? There could be trillions of small things that could become potential causes. Even your sneezing on a certain boring day, which you do not remember or feel guilty about, can be the cause of America's next big disaster. After the 1987 stock market crash, the 2008 financial crisis or the 2020 pandemic, many were forced to recognize that the future might never be predictable.

A good theory must be based on directly observable quantities, but this does not mean that only observables can be considered in a scientific theory. Einstein argued that it is quite wrong to develop a theory based on observation alone. In reality, the very opposite often happens: It is the theory – which might include unobservables – that decides what we can observe, and, consequently, observations can confirm the theory objectively. So, it seems important that we clarify the limits of both deductive and inductive logic because neither the rationalist nor empiricist approach alone will allow us to reach trustworthy knowledge with complete confidence.

2.1 Reasoning Solely Based on Induction

Based on observations or evidence alone, what can we say about general propositions? For example, consider the general proposition that all ravens are black. Saying that all the ravens you have observed are black is not enough to claim that this proposition is true, since there may be a non-black raven that has yet to be observed. Even if we can confirm that all the ravens that have existed up until now are black, does this guarantee that all the future ravens will be black as well? Regardless of how many black ravens we have observed, we cannot confirm with complete confidence the simple proposition that all ravens are black. Very disappointing, isn't it? How about the mortality of human beings? It is like the raven example above. No one can see the deaths of all human beings. Besides, is it not possible that humans will find a way to achieve immortality?

In response to Hume's agnosticism, Popper (1959) argued that falsification of the general proposition is the only option. It is grounded in his belief that the general proposition is not provable, so confirmation is impossible. He claimed that a conjecture-falsification cycle is the only way of constructing knowledge. However, falsification may not be as straightforward as it sounds. Kuhn (1962) argued that an established paradigm is resistant to falsification by a few pieces of negative evidence and can often be preserved by treating these instances as anomalies or by articulating ad hoc hypotheses.

FIGURE 2.1
Does the appearance of a white raven negate the general proposition 'All ravens are black'? Or, is it the exception that proves the rule?

Is the general proposition 'All ravens are black' proven false if a non-black raven is observed? Although white ravens have been observed in many parts of the world, we still have not officially falsified the proposition. Korea even has a legend about this white bird, which is considered an auspicious sign from heaven. More than a thousand years ago, ancient Koreans believed that a white raven was a sign of Korean unification sent from heaven. Regarded as an exception, the temporary appearance of a white raven did not falsify the general proposition.

A white raven appeared ten years ago in Korea. Today, no one is surprised by the appearance of a white raven. Through advances in biology, a white raven is easily explained along the same lines as a white tiger, a mutant that lacks melanin. Following this example, there are no general propositions that have no exceptions. Therefore, some may be content with saying that 'normal' ravens should be black. But what is the meaning of 'normal'? Does normal mean black here? Does this general proposition mean that ravens under normal conditions, excluding any abnormal situations, are black? What this implies is that under exceptional conditions, a raven can be non-black and this observation won't falsify the general proposition.

This shows that it is actually difficult to disprove a general proposition on the basis of observation alone. We still need an explanatory theory even for the falsification. For example, ravens need to have certain variants in a set of genes to have black feathers. We can explain why a particular raven is not black by showing what mutations are needed in the gene set. Popper insisted that a good scientific theory should be falsifiable, but it cannot be confirmed. However, despite being one of the greatest philosophers, Popper might be too extreme in his views. Some philosophers, such as Carnap, believe that we should be able to confirm a theory by means of induction. We should be able to change our mind following new evidence; we can go either from confirmation to falsification, or from falsification to confirmation.

Another difficulty in induction is illustrated by the example of black swans. Swans were believed to be white until someone found a black swan in Australia. Black swans are now used to represent unexpected or unimaginable extreme events and to illustrate the difficulty in predicting rare events. Since extreme and rare events play an important role in our lives, we should be able to explain their existence. For instance, suppose that we want to know the average height of people. A sample of about 100 people would be enough to have a reasonable guess. Even if we included the tallest man in the world, the sample average would not change much. However, if we wanted to know the average income of people, it would become a completely different story. If we had the richest man in the world in the sample, the average would certainly be much larger than the real average of the whole world. In areas like finance or insurance, it is important to prepare for extreme rare events.

In summary, the white raven is used to illustrate the difficulty in inductive confirmation, whereas the black swan is used to illustrate the difficulty in inductive prediction. But awareness of both can be helpful in promoting inductive reasoning.

2.2 Reasoning Solely Based on Deduction

Rational deductions can arrive at substantive conclusions by thinking alone, without appealing to experience, evidence or data. Can we answer the ultimate question whether God exists by thinking alone? In other words, can we find self-evident axioms that lead to the existence of God as the inevitable conclusion? Religious philosophers had proposed such arguments, but let's listen instead to some of the greatest mathematicians, notably René Descartes, Gottfried Leibniz and Kurt Gödel.

Descartes's deduction is similar to traditional philosophical arguments, which rely on a definition (D) that God is a being having all perfections, and an axiom (A) that existence is perfection. Therefore, God exists. Indicating the refined thinking of a mathematician, Leibniz was not happy with the traditional arguments since they implicitly assumed that God was possible, which for him required a proof. So, based on the same axiom, he instead proved two theorems: (i) God is possible, and (ii) if God is possible, then God exists (Lenzen, 2017).

Many commentators have questioned axiom A. A real advancement must wait until Gödel's proof, which dated back to the 1940s, but was only revealed in 1970, when Gödel thought he was dying. Reportedly, he did not want to publish it because he was afraid people might think he believed in God when, in fact, he was only interested in the proof as a logical exercise.

The proof relies on three definitions and five axioms. The axioms are technical symbolic statements in modal logic, so their self-evidence is difficult to judge; for instance, the first axiom is

$$(P(\phi) \wedge \Box x(\phi(x) \Rightarrow \psi(x))) \Rightarrow P(\psi).$$

To see if they are reasonable, the axioms roughly mean the following: (A1) Positive properties cannot imply negative properties. (A2) There are only positive and negative properties, and no neutral ones. (A3) Being Godlike is a positive property. (A4) Any positive property is the same in all possible worlds. (A5) Existence is a positive property. He proved three preliminary theorems and one final theorem that there is a Godlike object in every possible world. The proof has recently been computer-verified to be valid,[1] meaning that the existence of God does follow from the axioms.

How can one deny the existence of God? A few days after his historic flight on April 14, 1961, Yuri Gagarin, the first human to have gone into outer space, supposedly said, 'I looked and looked and looked but I didn't see God.' To this, C.S. Lewis, a British writer and lay theologian, responded, 'When a Russian cosmonaut returned from space and reported that he had not found God, this was like Hamlet going into the attic of his castle and looking for Shakespeare.' This shows the difficulty of falsification by means of induction. Actually, the authenticity of Gagarin's statement has been disputed. Colonel Valentin Petrov, a close friend of Gagarin, stated in 2006 that the cosmonaut never said such words and that the quote originated from Nikita Khrushchev's speech at the plenum of the Central Committee about the state's anti-religion campaign, saying 'Gagarin flew into space, but didn't see any god there.' This sounds reasonable, since Gagarin himself was a member of the Russian Orthodox Church. If a saint had travelled to space, he might have been able to find a sign of God. Thus, neither observation nor rational deduction alone provides a convincing justification or falsification of this proposition.

Both Descartes and Leibniz firmly believed in the existence of God, that is, $\Pr(G) = 1$ a priori in their minds, where the probability $\Pr(G)$ represents the degree of belief in the proposition G, whereas Khrushchev surely did not believe in God, or $\Pr(G) = 0$ a priori. We will show in Section 8.6 that it is impossible to change someone's mind with new evidence when they start with $\Pr(G) = 0$ or $\Pr(G) = 1$ a priori. At the end of the day, with Gagarin, we cannot be sure of what the cosmonaut really believed, since we do not know what he said exactly. In line with this, Kant asserted that due to the limitations of argumentation in the

[1]https://www.spiegel.de/international/germany/scientists-use-computer-to-mathematically-prove-goedel-god-theorem-a-928668.html

absence of irrefutable evidence, no one could really know whether there is a God or not. It is not a subject of deductive, logical or mathematical proof. Rather, it may be considered as a creed, closer to a matter of belief than knowledge. Nevertheless, according to Kant, it needs to be true for the purpose of human moral laws.

2.3 Complementary Induction and Deduction

In modern science, the two modes of reasoning are not seen as competing alternatives, but as complements to each other. We present them schematically in Figure 2.2. All the evidence (facts, data, observations) comes from the real world, as can be seen on the left-hand side of the figure. We mark the evidence with rectangles, representing the observables whose reality can be comprehended by our brain.

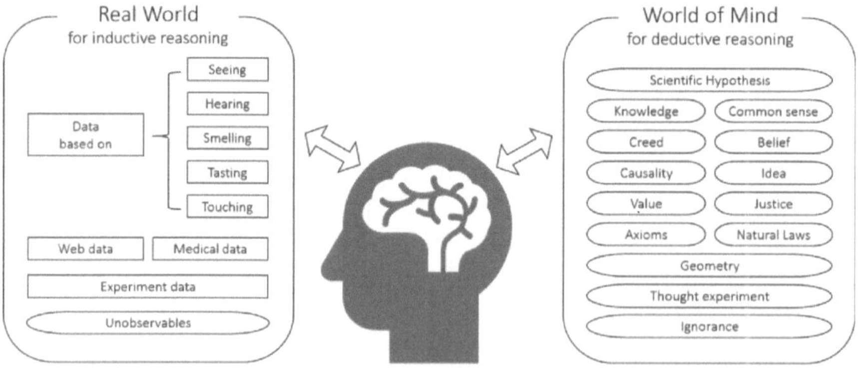

FIGURE 2.2
The elements of inductive reasoning (left-hand side) and deductive reasoning (right-hand side).

In the real world, we can only learn about things outside ourselves by using our senses and our instruments. There are many things that we cannot observe even with the help of current instruments. Natural laws cannot be observed, but physicists believe their laws are real, perhaps as Kant believed that his moral laws were real in this world, or as Plato thought that his idea world was real and regarded all objects in this world as just an imitation of his idea. There are also things that we cannot observe directly but whose existence can be inferred.

In physics, numerous attempts to observe dark matter have not been successful. Dark matter is supposed to constitute 85% of the total mass of the universe, while dark energy plus dark matter together constitutes 95% of the total mass-energy content. Although its existence is generally accepted by the scientific community, dark matter has yet to be detected, allowing us to conclude that if it exists, it must barely interact with ordinary matter and radiation. Thus, dark matter is currently placed in the world of mind, and will be placed in the real world if observed later. But there are physicists who want to regard dark matter as unobservable in the real world. So, even in physics, the unobservable is necessary to explain the universe.

Thus, in Figure 2.2, the real world is composed of observables (inside the rectangles) and unobservables (inside the ovals). Science explains or predicts the behaviour of observables, typically via unobservable laws. Induction is the means of finding explanations for the interactions between the real world and our brain. We can check observables by comparing what is observed in the real world with what we predict in our minds. If the observed results turn out the way we expected, we can say that we have a convincing theory to explain the real world. In the real world, we can conduct experiments or make observations to verify our theory, which is like receiving a Supreme Court decision on our scientific theory.

There is an epistemological world, where we interpret, think and explain how the world operates via our brain. Our knowledge, propositions, scientific theories, and values belong primarily to this world. Plato thought that this 'idea world' was more real, and that all the observables in the real world were but imageries, imitations of the real. In this world of thinking, rational deduction is the accepted logic. Thought experiments can be performed to check our thoughts (inductive reasoning in this deductive world of mind) as an alternative to inductive reasoning in the real world. However, our knowledge can be arbitrary if it cannot be confirmed by evidence from the real world. Consequently, uncheckable thought experiments need to be verified by experiments or observations in the real world. Einstein developed his general relativity theory via thought experiments, but his theory was confirmed by observations in the real world. As both inductive and deductive reasoning have their own drawbacks, they need to function as complements to overcome their weaknesses.

The way we comprehend an apple in front of us is a product of the brain's processing. Although it is not clear if others see the apple the same way we do, we still like to believe that humans, regardless of their sex, race, culture and generation, see objects in the same way. Thus, consensus knowledge refers to what all human beings accept as true. When unobservable laws seem to do a good job of explaining the real world, we tend to believe that such laws govern the real world. However, according to some philosophers such as Popper all universal laws are false – strictly speaking, they have zero probability (see Popper (1959, Appendix vii) and also Section 8.6); such laws are only temporarily accepted until they are falsified. Unobservables in the real world could be a source of ignorance in the world of mind because we cannot verify them unless they interact with what we can observe. Nevertheless, great discoveries in science always find unobservable laws that explain observables in the real world. Justification of unobservables via deductive logic without observational evidence is allowed in meta-science, which effectively highlights the difference between science and meta-science.

The painting in Figure 2.3, by the Italian Renaissance artist Raphael, depicts an older Plato walking alongside a younger Aristotle. Plato points

FIGURE 2.3
An older Plato and a younger Aristotle in Raphael's painting The School of Athens.

up, representing the world of mind, while Aristotle points down, representing the real world. Plato believed that the world of the mind was real and that the observables in the real world were its imitations, whereas Aristotle thought that the observables in the real world were simply real. Kant connected the two by claiming that both were the works of our brain. That is, we see the real world as our brain shows it to us and we think as our brain leads us to think. Thus, our knowledge, the product of our brain, requires both inductive and deductive logic. In modern science, deductive reasoning is confirmed by inductive evidence, and we say that it actually describes the real world.

2.4 Abduction

Even though much of our daily reasoning is inductive, it is tricky and we do not necessarily get better at it over time. The reasoning is based on incomplete information and produces uncertain statements, so it is easy to fall into traps. Perhaps highlighting its trickiness, some philosophers proposed a third mode of reasoning called 'abduction,' which is supposedly distinct from deduction and induction. Sometimes it is known as 'inference to the best explanation,' and also closely related to Polya's plausible reasoning (Polya, 1954). A closer analysis, however, shows there is no need for a distinct third mode: it's much simpler and more rigorous to put abduction within the inductive mode of thinking of formal statistical inference. In Section 8.6 we will describe plausible reasoning based on a qualitative Bayesian theory. Here is an example:

- **Deduction:** All men are mortal; Socrates is a man. Therefore, Socrates is mortal.
- **Induction:** Socrates is a man; he is found to be mortal. Therefore, all men are mortal.
- **Abduction:** All men are mortal; Socrates is mortal. Therefore, Socrates is probably a man.

You may immediately raise objections to the simplistic example, but it does illustrate the logical direction between the general and the specific statements, and the nature of the conclusions. In induction, the conclusion is a general statement, but in abduction it refers to a specific case stated in the premise. This is the kind of reasoning that doctors use in medical diagnosis or detectives use to solve criminal cases. Suppose that John has a collection of symptoms that are known to be caused, say, by colon cancer. Then, by abductive reasoning, he *probably* has colon cancer. Classical inductive reasoning does not apply, as we are not interested in making general statements about men or colon cancer, but only in diagnosing John's illness.

A version of abductive reasoning can also be seen in a generic scientific process, where one proposes and tests a theory. This also fits the Popperian program of falsifying a hypothesis. Schematically, theory A implies a testable prediction B. One performs an experiment based on A designed to produce B. If B is indeed observed, then by abductive reasoning, the scientist would be very happy that A is verified and probably correct. Certainly no one can say that A is proven, since mathematically, from the assumption that A implies B, observing B tells us nothing about A.

However, in science, it is commonly accepted that evidence B supports theory A. The weakness of this line of argument is that there is no formal indication of the kind of testable prediction B provides a good test for A. Mathematically, assuming A implies B tells us nothing about what happens to B if A is wrong. But from experience, good scientists intuitively know how to choose experiments that provide a good challenge to the theory, in the sense that if A is wrong, then B is highly unlikely to occur. If B indeed does not occur, they have to modify or even kill the theory, but observing B would be highly indicative of the truth of A.

When Einstein proposed the general theory of relativity in 1915, one of his astonishing predictions was that space is curved by masses, which in turn explains gravity. His first major test of the theory to explain the so-called perihelion motion of Mercury, a slight shift in the direction of its closest radius from the Sun. The shift was previously known before Einstein's theory, but there is a discrepancy between the observed shift (9.55 arc minutes/century) and the predicted value using Newton's law of gravitation (8.85 arc minutes/century). Einstein was justifiably elated by the success of his theory, as intuitively it is impossible for a wrong theory to come up with a beautiful mathematical formula that gives just the right amount of correction (0.7 arc minutes/century) to Newton's prediction.

Around 1867, the American polymath Charles Peirce first introduced 'abductive reasoning' – which, confusingly, he called 'hypothesis' – as a distinct mode from induction or deduction. He published a book on probable inference in 1883. Yet, in 1910, Peirce admitted that 'in almost everything I printed before the beginning of this century, I more or less mixed up hypothesis [= abduction] and induction'! Unfortunately, the confusion seems to continue to the present, where we can still see many articles, books or workshops about abductive reasoning as a distinct mode of reasoning. As we comment above, it's simpler to put abductive reasoning within the context of formal statistical reasoning.

2.5 Category-Based Induction

Classical examples of induction are as follows: we see evidence of one or more black ravens, so we conclude – or, more correctly, we conjecture – that all ravens are black. Statistics has grown to make this reasoning quantitative and rigorous. However, indicating the richness of inductive reasoning, evidence can be in the form of categories rather than just individuals. Interestingly, some category-based induction involves reasoning that is not easily covered by standard statistical methods. We have included this topic since, as illustrated later, the phenomena have general relevance in qualitative scientific and statistical thinking.

In a seminal paper, Osherson et al. (1990) provided an empirical and theoretical study of a compendium of induction 'phenomena,' involving arguments where the premises and conclusion are of the form 'All members of category C have property P.' The generic form of an argument is

$$\frac{\text{Grizzly bears love onions.}\quad \text{Polar bears love onions.}}{\text{All bears love onions.}} \qquad \text{(A1)}$$

All bears, and grizzly and polar bears are categories. Mimicking the terminology of deduction reasoning, Osherson et al. call the statements above the line the premises, and the one below the line the conclusion. Argument (A1) looks like a standard induction, where we can think of the premises as the 'evidence' to support the conclusion. However,

Osherson et al. also consider arguments with specific conclusions, such as

Robins use serotonin as a neurotransmitter.
Blue jays use serotonin as a neurotransmitter.
―――――――――――――――――――――――――
Sparrows use serotonin as a neurotransmitter. (A2)

One can compare the qualitative strength of arguments, where strength measures how much the belief in the conclusion follows from the belief in the premises. Prior belief in the conclusion should not be taken into account. For instance, (A2) is judged stronger than the following argument:

Robins use serotonin as a neurotransmitter.
Blue jays use serotonin as a neurotransmitter.
―――――――――――――――――――――――――
Geese use serotonin as a neurotransmitter. (A3)

This is because robins and blue jays are judged to be more similar to sparrows than to geese. What kind of general reasoning is used in category-based induction? For (A2), one might think that the reasoning goes first to the general and then back to the specific. That is, given the premises about robins and blue jays, perhaps all birds use serotonin as a neurotransmitter; if so, then sparrows do too. But it's not that simple, since for the geese, the argument is not judged to be as strong.

The investigations of Osherson et al. suggest some expected and some rather unexpected phenomena, most of which have previously appeared in the literature. These phenomena can be seen as informal principles rather than formal rules. It's useful to distinguish between general and specific arguments, since the type affects reasoning:

General: The premise categories are members of the conclusion category; for instance, (A1) is general.

Specific: If any of the premise and conclusion categories belong to a higher-order category C, the other categories must also belong to C. For instance, (A2) is specific.

The general and specific arguments share some properties, e.g., diversity and monotonicity of premises increase strength. The effect of diversity is perhaps a bit unexpected; see below. Monotonicity means increasing the number of premises of similar categories, so the strengthening effect is expected. There are also mixed arguments, which we will not cover. Here is a selection of the phenomena for different types of arguments. Most claims of the argument strength were validated empirically by Osherson et al.

1. *Premise typicality – general:* The more typical the premise categories are in the conclusion category, the stronger the argument. For instance,

   Robins use serotonin as a neurotransmitter.
   ___
   Birds use serotonin as a neurotransmitter.

   is judged stronger than

   Penguins use serotonin as a neurotransmitter.
   ___
   Birds use serotonin as a neurotransmitter.

   This is because robins are more typical of birds than penguins are. As a simple application, medical results obtained from a more representative sample will always be judged stronger than results obtained from a highly selected group:

   A sample of individuals from the general population responds to non-steroid anti-inflammatory drugs.
   ___
   All humans respond to non-steroid anti-inflammatory drugs.

   is stronger than

   Professional athletes respond to non-steroid anti-inflammatory drugs.
   ___
   All humans respond to non-steroid anti-inflammatory drugs.

2. *Premise diversity – general:* The more diverse the premise categories are among themselves, the stronger the argument. For instance,

   Robins use serotonin as a neurotransmitter.
   Penguins use serotonin as a neurotransmitter.
   ___
   Birds use serotonin as a neurotransmitter.

is stronger than

$$\frac{\text{Robins use serotonin as a neurotransmitter.} \quad \text{Blue jays use serotonin as a neurotransmitter.}}{\text{Birds use serotonin as a neurotransmitter.}}$$

An important application of this principle is in validation studies. Suppose that we have discovered something interesting in a dataset. How do we choose the validation set in order to get a strong overall conclusion? The strength affects, for example, its generalizability. The principle would say that the more distinct the validation dataset is from the discovery set, the stronger the conclusion. A discovery in men becomes more compelling if validated in women than if validated in another group of men. When the validation and discovery sets are too similar, such as when they are the result of a random split of a larger dataset, the validation is weakest. Non-random separation of validation and discovery datasets, in general, produces a stronger conclusion. This validation property is perhaps more appreciated by scientists than by statisticians.

3. *Conclusion specificity – general:* There is always some hierarchy of categories. The more specific the conclusion, the stronger the argument.

$$\frac{\text{Lung cancers are curable by cisplatin.}}{\text{Solid cancers are curable by cisplatin.}}$$

is stronger than

$$\frac{\text{Lung cancers are curable by cisplatin.}}{\text{All cancers are curable by cisplatin.}}$$

This phenomenon is to be expected as it is a version of the first principle on typicality. Lung cancers are solid cancers, while all cancers include blood cancers, which have strong differences from solid cancers.

4. *Premise-conclusion similarity – specific:* The more similar the premise categories to the conclusion categories, the stronger the

argument. Examples are given by (A2) and (A3). To take a medical example,

$$\frac{\text{mRNA vaccines for covid-19 are effective on monkeys.}}{\text{mRNA vaccines for covid-19 work on humans.}}$$

is stronger than

$$\frac{\text{mRNA vaccines for covid-19 are effective on mice.}}{\text{mRNA vaccines for covid-19 work on humans.}}$$

5. *Premise diversity – specific:* Similar to No. 2 for a specific argument, more diverse premises bring stronger support.

$$\frac{\begin{array}{c}\text{Lions use norepinephrine as a neurotransmitter.}\\ \text{Giraffes use norepinephrine as a neurotransmitter.}\end{array}}{\text{Rabbits use norepinephrine as a neurotransmitter.}}$$

is stronger than

$$\frac{\begin{array}{c}\text{Lions use norepinephrine as a neurotransmitter.}\\ \text{Tigers use norepinephrine as a neurotransmitter.}\end{array}}{\text{Rabbits use norepinephrine as a neurotransmitter}}$$

This is quite unexpected, as giraffes are not more similar to rabbits than tigers are. Similar reasoning would apply in validation studies, where one should try to validate a scientific result in more diverse populations or situations.

Some theory of category-based induction

It's clear from the examples, that we cannot apply standard statistical theories to establish the strength of an argument, as the premises may not resemble a random sample from any population at all. Osherson et al. developed a descriptive and qualitative theory based on two factors: (i) the similarity between the premise categories and the conclusion category; and (ii) the similarity between the premise categories and the members of the lowest-level category that includes both the premise and conclusion categories.

The first factor is perhaps obvious, and its effect can be seen in many examples. The second factor is less so. For instance, in (A2) the lowest category that includes robins, blue jays and sparrows is the order Passeriformes – perching birds. But in (A3), we have to go all the way to the bird class to include robins, blue jays and geese. Osherson et al. theorize that when judging arguments we take a linear combination between these two factors, with individualized weights, so different individuals may arrive at different judgements. We won't cover any further details of the theory here, as they are not as illuminating as the examples.

3

Hilbert's Broken Dream: Limitations of Deductive Reasoning

Mathematics, 'the queen of the sciences,' is often seen as a way of obtaining absolutely objective, true knowledge with complete certainty. These features of mathematical truths, particularly in geometry and arithmetic, have always impressed philosophers. They realized that the certainty comes from deductive reasoning starting with self-evident axioms. As Leibniz hoped, with the right guiding principles, we 'should be able to reason in metaphysics and morals in much the same way as in geometry and analysis.' Kant elevated the notion of self-evident axioms to synthetic a priori knowledge, to be utilized in a similar way. Only in the 20th century did we realize that even mathematics is not certain, that certainty is only a matter of definition.

In mathematics and logic, the truth of a mathematical proposition H is an indicator function

$$I(H),$$

which takes only the integer value 1 (true) or 0 (false). Rationalists such as Leibniz or Hilbert dreamed of inventing an operational machine that produces truth values for all mathematical or logical propositions. They wished to have an automatic way of knowing the true value of $I(H)$ and to generate all mathematically true propositions by a machine. Towards the end of the 19th century, a group of prominent mathematicians and logicians such as Hilbert and Giuseppe Peano had the ambition to embark on such a project. The purpose was to find a procedure such that all mathematical propositions could be mechanically and automatically deduced by strict irrefutable logic from the rightly chosen axioms.

If the procedure were implemented in a machine, then we could theoretically ask the machine to find all true propositions, even those for which we did not have proof yet. In this scenario, humans would have perfect knowledge, at least in the mathematical wonderland run by the axioms. The idea was attractive and influential, encouraging, for example, Whitehead and Russell to write *Principia Mathematica* (1910) on the foundations of mathematics using logic. Similarly, the foundations of ethics, statistics, justice, etc. have been proposed, so we can fairly understand and describe these fields, provided that they are governed by their presumed basic propositions.

Suppose that we have a proposition H that is not yet known to be true or false. What mathematicians and logicians dreamed was that

$$\Pr(H) = I(H).$$

This means that what we believe in the world of mind, represented by $\Pr(H)$, would be true in the real world, $I(H)$. Then, at least in mathematics, they would have access to the whole truth. It seems natural that science should also follow mathematics to attain the whole truth of the universe. The theme of the Enlightenment is that with the triumphant advancements in science, it is possible to predict the future and all the phenomena. The so-called Laplace demon, the belief in universal determinism, stems from the belief in Newtonian laws. Up until the early 20th century, great mathematicians tried to find a way to access the whole truth, at least in mathematics. But it turned out to be a mission impossible.

3.1 Is Euclidean Geometry True?

To the Greeks, absolute truth, if it exists, is never changing, and such truth can only be obtained by mathematics. Thus, they believed that it is possible for humans to access the truth. Euclidean geometry enjoyed a special status as the beacon of mathematical truth. Euclid showed that, based on several self-evident axioms, all propositions in geometry could be proven to be absolutely true, without uncertainty and without any external evidence. Plato founded his Academy in Athens in 387 BC, where he stressed that mathematics was a way of understanding universal reality. He believed that geometry was the key to unlocking the secrets of the universe. So much so that at the entrance to his Academy, he posted the sign: 'Let no one ignorant of geometry enter here.'

FIGURE 3.1
Raphael's painting The School of Athens, completed in 1511, depicting many ancient Greek philosophers and mathematicians.

Euclidean geometry as a paradigm for the world of geometry is built up from five axioms. The fifth postulate, also known as the parallel postulate, states:

> If a straight line falling on two straight lines makes the interior angles on the same side of it taken together less than two right angles, then the two straight lines, if produced indefinitely, meet on that side on which the sum of angles is less than two right angles.

That looks a bit inelegant, so the axiom became a target of suspicion that it was not necessary. But within the context of Euclidean geometry, it is equivalent to the Playfair axiom:

> In a plane, through any given point P not on a line L there passes exactly one line parallel to L.

In this form, the axiom certainly looks so intuitively obvious that no one questioned it for more than two millennia, during which Euclidean geometry maintained its status as the normal science of geometry. Everyone believed $I(H) = 1$ without any doubt. If you don't believe it, go ahead and try to plot a parallel line passing a specific point in the paper. Even children will immediately understand it. Kant regarded Euclid's axioms as a prime example of synthetic a priori knowledge.

FIGURE 3.2
Euclid in Raphael's painting The School of Athens.

A great challenge in setting up an axiomatic system is to show that the axioms are necessary and sufficient. Sufficiency means that the stated theorems are proved using only those axioms, i.e., no hidden axioms. Each axiom should be necessary in the sense that there is no redundancy; in other words, they are all mutually independent, or no axiom is a consequence of the others. (Ideally the system is complete, which we will discuss below in connection to Gödel's theorems.)

It is not well known that Euclid's system is not sufficient to prove some standard theorems of geometry; a revised system is provided, for instance, by Hilbert. However, historically, the suspicion of the non-redundancy property was widely known: There were many attempts to prove the parallel postulate using Euclid's other axioms. But all such attempts had failed. There is even a story of a man who spent endless nights in this fruitless pursuit. He finally gave up and wrote a death will to his son, imploring him not to pursue this matter further. But his son, Janos Bolyai, continued the quest and finally concluded that the parallel axiom was independent of the other axioms. If an axiom is a consequence of the other axioms, then negating it will produce an inconsistency and contradictions. Bolyai did not find any contradiction by negating the parallel axiom; what he found instead were other consistent geometries.

It is also well known that the great mathematician Gauss discovered a postulate that could replace the parallel postulate and give birth to non-Euclidean geometry, but in fear of controversy, he hid it inside the drawer of his desk. (So even Gauss acquiesced to the prevailing Euclidean paradigm.) Fortunately, around 1830, Lobachevsky and Bolyai finally negated Euclid's parallel postulate and created a new non-Euclidean geometry, namely hyperbolic geometry. Poincare's upper half-plane model serves as one of the examples of hyperbolic geometry. In this model, straight lines are either vertical lines or half circles in the upper half of the complex plane. Hence, parallel lines can meet in hyperbolic geometry.

In the middle of the 19th century, Gauss-Lobachevsky-Bolyai's non-Euclidean geometry was further systematically developed by Gauss's student, Bernhard Riemann, under the name Riemannian geometry. Riemannian geometry includes non-Euclidean geometries such as elliptic and hyperbolic geometry. One of the well-known theorems in Euclidean geometry is that the total sum of the angles in a triangle is 180 degrees. However, in elliptic geometry and hyperbolic geometry, it can be greater than or less than 180 degrees. The truth of Euclidean geometry can be checked by measuring the exact angles of a triangle. Thus, only experience (observation, evidence), and not deductive reasoning, can test Euclid's system.

We now know that the Earth resembles a sphere rather than a flat plane. Suppose that we want to construct a geometric system on the sphere, on which a straight line makes a great circle. On the sphere, every pair of distinct great circles intersects twice. This means that there are no parallel lines on the sphere that do not cross each other twice. Thus, on the sphere, the parallel postulate cannot hold. In other words, on the Earth's surface $I(A) = 0$.

So, the parallel postulate is indeed not necessary for building a geometrical system. It holds true only in a flat world, but didn't we believe we lived on a flat earth until Magellan circumnavigated the earth? So it is relatively recent that we got the evidence that the earth is not flat but spherical. There are, of course, still members of the Flat Earth Society, who seriously believe and claim that the earth is flat. What would a flat earth look like? Figure 3.3 shows a flat-earth map with the North Pole in the centre of the world and a ring of Antarctic ice surrounding the edge of our planet.

After Einstein, we believe that non-Euclidean geometry is a better way of representing what space-time actually looks like. Imagining the geometry of the universe is a great mental experiment. But how can we actually

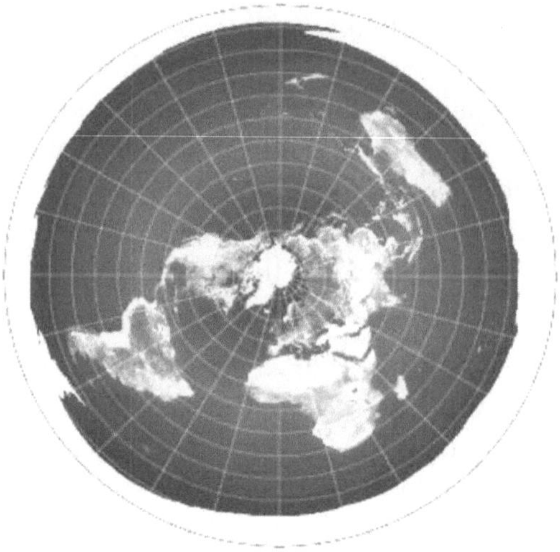

FIGURE 3.3
A possible map of flat earth by Strebe; Creative Commons Attribution-Share Alike 3.0 Unported license.

know this since no one has ever travelled to the galaxy? All the apparent facts that we believe in our small world might not be necessarily true in an outer, larger world (galaxy) that has not been explored yet. How can we know that the axioms of our imagination (mind) established by our brain are true? It may be a Platonic ideal to believe that we can access absolute truth solely through rational deduction if the axioms are uncheckable.

3.2 Gödel's Incompleteness Theorems

Rationalists such as Leibniz and Hilbert dreamt of finding a way to automatically calculate the truth values of all mathematical propositions. Can we enumerate all true propositions in mathematics just as a machine produces sausages? It's a machine, Poincare skeptically remarked in 1908, 'where we should put in axioms at one end and take out theorems at the other, like that legendary machine in Chicago where pigs go in alive and come out transformed into hams and sausages.'

In 1931, Gödel published groundbreaking works on mathematical logic that highlight the limitations of formal axiomatic systems. Say, someone proved a certain proposition using a set of axioms, and someone else disproved it by using the same set of axioms. Then, we can say that the system of axioms is inconsistent, as it can produce a contradiction. Therefore, to establish any mathematical proof, we have to start with a consistent set of axioms. But the set must ideally be complete in the sense that *all* logically meaningful propositions within the system must be provable or decidable. Gödel's first incompleteness theorem states that any consistent system of axioms is incomplete: there are propositions that can neither be proved nor disproved within the system. Thus Gödel showed that Hilbert's dream is indeed only a dream. (Formally, Gödel's theorems apply to axiomatic systems rich enough to cover basic arithmetic, but informal interpretations that do not mention the technical assumption are more striking and attractive.)

One famous example of an undecidable statement is the continuum hypothesis in set theory, which states that there is no set whose cardinality is strictly between that of the integers and that of the real numbers. This hypothesis was the first of Hilbert's 23 problems presented at the International Congress of Mathematicians in 1900. In set theory, the Zermelo-Fraenkel (Z-F) theory was proposed to obtain a set theory that was free of paradoxes. In 1940, Gödel showed that the negation of the continuum hypothesis is consistent with the Z-F axioms. Later in 1963, the American mathematician Paul Cohen proved that the continuum hypothesis cannot be derived from the Z-F theory together with the Axiom of Choice, thus establishing the independence of the continuum hypothesis. (Thanks to this groundbreaking work, in 1966 Cohen was awarded the Fields medal, also known as the Nobel Prize in mathematics.)

An obvious interpretation of Gödel's theorem is that there are some propositions whose truth we will never know. After all, there are so many more truths out there than just the number of comprehensible or provable propositions. As such, some truths may never be in our grasp. In deductive logic, including mathematics, there is always uncertainty that cannot be eliminated.

So, what do we do with the continuum hypothesis? We could add other axiom(s) that would lead to a proof of the continuum hypothesis. Alternatively, we can add the hypothesis itself to the set of axioms. This would clearly imply that the axiom is true because we simply declare it to be so. Thus we arrive at the logical-positivist position that mathematical truth is only analytic, i.e., propositions – even axioms – are only true by definition, not for any innate reasons, and certainly not synthetic a priori as claimed by Kant.

Another wake-up call from Hilbert's dream was provided by Alan Turing, the legendary English polymath. He reformulated Gödel's 1931 results, replacing the formal language based on arithmetic with a hypothetical device known as the Turing machine. The machine is a universal machine capable of performing any conceivable mathematical computation; it is practically realized as a modern computer.

Given Gödel's 1931 results, one might ask a similar question if we can compute by means of a computer whether any proposition can be proven or not. Turing proved that the answer is negative. He showed that it is not possible to know whether the Turing machine will ever halt or not. It is like giving a robot manoeuvring a spaceship a mission to find Utopia in galaxies. Turing said that it is impossible to know whether the robot will come back in finite time to give an answer (halting) or will continue exploring the space (no halting). We cannot even calculate whether a computer will return an answer during our lifetime or not. Therefore, we can say that uncertainty is unavoidable even in a computational, algorithmic world.

Leibniz, Hilbert and Gödel all employed the expression 'automatically.' This means that a machine can be built such that it automatically produces true propositions, like a machine that makes German sausages. Turing's great contribution is that he actually defined what it means to do something automatically. This led to the development of modern computers and truly changed the course of humanity. Turing showed that all automatic deductive logic can be done by his Turing machine. Ironically, this great gift to mankind embodies the limitations of our own deductive logic. Its use of automatic learning based on inductive reasoning led to a new era of artificial intelligence (AI), where the limitations of deductive logic are remedied by inductive logic.

Gödel's theorems and arbitrariness of deduction

Suppose that we have developed and worked with a system of axioms for a while and never found any contradictions. We'd be tempted to declare the system is consistent. But, can we prove it so that we are 100% certain? Gödel's second incompleteness theorem says: No, the consistency of a consistent system of axioms cannot be deductively verified by itself. That is, it is not possible to prove internally that the system is free of contradictions. Fisher, the 20th-century great British statistician and geneticist, had a clear intuitive explanation for this theorem (Fisher, 1958). We know that, if a basic proposition G is false, conclusions driven by deductive logic based on G being true are arbitrary. So, with a wrong axiom, we can prove any proposition. For instance, suppose that someone believes that 2+2=5. Based on this axiom, we can prove any proposition, such as 'Linda – or anyone you know – is the Pope.' If 2+2=5, then 5=4. If we subtract 3, we will find 2=1. This means that Linda and the Pope, two different people, are the same person. Therefore, Linda is indeed the Pope.

Fisher continued: Suppose a Ph.D. student came, breathless with excitement, and said, 'I have proven that this system of axioms is free from all contradictions.' You'd say, 'Did you prove it using only those axioms?' He might say, 'Yes, I have written out a chain of propositions that demonstrate that these axioms are free from contradictions.' Well, you'd look at him with mild surprise and say, 'I suppose you know that if this system of axioms did contain a contradiction, you could prove exactly those same propositions!'

Taken together, Gödel's theorems say that in a formal deductive system there are statements whose truth cannot be decided; we cannot verify the consistency of a system of axioms by itself; and there are true statements that cannot be proven within the system. Therefore, consequences or conclusions solely based on deduction can be uncertain and arbitrary. When we encounter unprovable statements, we are forced to rely on other sources of information, such as intuition, empirical evidence, or meta-mathematical arguments, in order to determine their truth value. This means that our acceptance or rejection of such statements can be arbitrary, since we may be relying on non-deductive methods that are not completely reliable or objective.

We have also seen that in geometry we can have different sets of axioms in Euclidean and non-Euclidean geometries. So, in mathematics, we have certainty at the price of arbitrariness; there is no absolute truth, only truth relative to the set of axioms. In general, a system of axioms would constitute a model, which may or may not fit reality. If, as the criterion for scientific truth, we require the model to agree with reality, then experience – processed with evidence-based induction – is crucial for validating the truth of a mathematical logico-deductive system. To a large extent, that's actually the attitude that statisticians and scientists have about their models and their theories. Although we do not necessarily believe the models to be true, they are useful for describing and understanding reality or for deriving predictions. As the statistician George Box famously said, 'All models are wrong, but some are useful.'

4

'Real' Scientific Process

4.1 Theory vs Observations

What comes first: theory or observation? With induction, one starts with observations to speculate on or conjecture general theories. However, some philosophers such as Kant or Kuhn believe it's the other way around: that thoughts precede evidence; that a priori mental concepts are the organizing principles of our senses; that observations are 'theory laden,' etc. However, because of the constant interplay between thinking and observing, it is like asking what came first, the chicken or the egg.

The spirit of exploration in the Age of Enlightenment started a new era of scientific revolution: Galileo and Kepler proposed the view that Earth circles the Sun, and Isaac Newton explained how the fall of an apple and the planetary motions are due to gravity. Newton's theory, however, is not considered an explanation of gravity, since it does not specify any mechanism and it requires action at a distance. It took more than two centuries for scientists to find a real explanation. Let's review the process. In the 19th century, mathematicians realized that the parallel postulate may not be necessary in geometry. This led to the development of Riemannian geometry, which provided the basis for Einstein's general theory of relativity, which offers an explanation of gravitation as the effect of a local curvature in space-time. How do we weigh the relative contributions of theory vs observations here?

Checkability is important for both empiricists and rationalists. For a scientific theory, confirmation is carried out by comparing what the theory predicts in the world of mind with what we actually observe in the real world. Claudius Ptolemy, an ancient Alexandrian mathematician-astronomer-philosopher, started with the assumption that the Earth was the central body and demonstrated the orbits of the planets. This enabled him to predict eclipses of the Sun and Moon so precisely that his geocentric model was considered the foundation of astronomy for nearly 1500 years. However, it was eventually replaced by the Newtonian model. Although Ptolemy's predictions were often precise, they required a lot of modifications to achieve accuracy. Furthermore, with a wider scope of applications beyond astronomy, Newtonian physics is accepted as a better and more elegant theory in terms of both deductive and inductive reasoning.

In hard sciences like physics, there is a 'Supreme Court' on which everyone agrees and to which all disputes are taken. The court's decisions are final and are based on evidence. In the late 16th century, Galileo showed that instead of arguing about how objects should move according to philosophical or theological precautions or quoting ancient authorities such as Aristotle, it is better to simply use the evidence of our own eyes, just as his telescope confirmed his theory. In contrast, in the social sciences, where thinking subjects are studied, the related evidence is not very clear. Perceptions or observations play an important role in the establishment and confirmation of knowledge.

4.2 Paradigm Shift

The nature of scientific revolutions has been described by Kant, who used the phrase 'revolution of the way of thinking' to refer to Greek mathematics and Newtonian physics. In his *Structure of Scientific Revolutions,* Kuhn (1962) rephrased it as a 'paradigm shift.' Kuhn's novel approach as a philosopher is to describe science as it is practiced by fallible human scientists, not to use it as a seemingly perfect model for (re-)building philosophy, and not to provide prescriptions about how it should be conducted. The real world of science turns out to have some special characteristics, which are perhaps not as completely logical or rational as those presented by the rationalists but can be understood as the effort to balance between the need for stability and new ideas.

Scientists are natural skeptics. The strongest skepticism is of course directed towards what they consider non-scientific, but the skepticism may turn into conservatism when it comes to changing their prevailing theory. That is, they tend to be skeptical of, or even hostile to, new theories. When Alfred Wagener proposed the continental drift theory in 1912, he was met with significant resistance and even ridicule from prominent geophysicists, who believed that the continents were fixed. Even 'our' Harold Jeffreys, a notable geophysicist as well as probabilist whose name appears on many pages of this book, maintained his opposition to the theory until his death in 1989, despite its widespread acceptance in the 1960s.

One contributing factor to the resistance from geophysicists was Alfred Wegener's original background as a meteorologist rather than a traditional geophysicist. This is another human aspect of science/scientists: you don't like it if non-experts barge in like unwelcome guests and teach you something fundamentally new about your area of science. Only after an accumulation of evidence did the geophysicists accept the validity of Wegener's ideas, and his theory became an integral part of the broader concept of plate tectonics. This paradigm shift revolutionized our understanding of Earth's geology and continues to be a fundamental framework in geophysics and related fields today.

Kuhn defined 'normal science' as scientific work that is done within a prevailing framework. Paradigm shifts arise when the dominant paradigm under which normal science operates is rendered incompatible with new phenomena, facilitating the adoption of a new theory. He explained the development of paradigm shifts in science in four stages:

Normal science. In this stage, a dominant paradigm is active, characterized by a set of theories and ideas (axioms, basic propositions) that define what is possible and rational; for example, Euclidean geometry or Newtonian physics. A paradigm gives a view of how the world (subject of interest) operates under the laws that form normal science. Given the axioms, normal science is basically deductive logic. Notwithstanding Popper's view on falsification, *anomalies do not bring normal science down immediately.* It takes time for anomalies to accumulate and invoke inductive reasoning in some individuals to question normal science. This slow process of science may look irrational from the outside, but, like other humans with their tools, scientists do not like to easily replace theories that have served them well.

Extraordinary research. When enough significant evidence has been gained against the current paradigm, normal science is thrown into a state of crisis. To address the crisis, scientists push the boundaries of normal science in what Kuhn calls 'extraordinary research,' which is characterized by its exploratory nature. Without the structure of the dominant paradigm to depend on, scientists engaged in extraordinary research must produce new theories, thought experiments, and experiments to explain the anomalies. Kuhn sees activities during this stage – 'the proliferation of competing articulations, the willingness to try anything, the expression of explicit discontent, the recourse to philosophy and the debate over fundamentals' – as even more important to science than paradigm shifts.

Adoption of a new paradigm. According to Max Planck, a 20th-century German theoretical physicist, 'a new scientific truth does not triumph by convincing its opponents and making them see the light, but rather because its opponents eventually die out and a new generation grows up that is familiar with it.' A new paradigm needs new, young followers who can compete with and outnumber followers of old normal science. Paradigms are difficult to change, but new paradigms can gain influence by explaining or predicting phenomena much better than the previous paradigms or by being more subjectively pleasing. During this phase, proponents of competing paradigms address what rests at the core of the debate: Which paradigm will be a good guide for future problems that neither paradigm is capable of solving at the moment?

Result of the scientific revolution. In the long run, the new paradigm becomes institutionalized as the dominant one. Textbooks are written, obscuring the revolutionary process. With new axioms, it becomes normal science with its own deductive logic.

In summary, each paradigm constitutes a world characterized by basic propositions. Normal science is very fruitful in describing the world through deductive reasoning until anomalies arise. This is where evidence-based inductive reasoning comes into play to help guide a paradigm shift. There are disciplines that behave like religions, where evidence is not effective in bringing a paradigm shift, and the shift may not be clear or may not even exist. Throughout history, we have witnessed how what was once heresy sometimes becomes accepted normal science. If induction alone is not able to determine one paradigm with certainty, there could be multiple paradigms that can exist at the same time. In other words, while evidence-based falsification plays an important role in choosing a paradigm, multiple paradigms can coexist if falsification is impossible. We see this situation, for instance, in physics, where quantum gravity and string theories coexist without a clear winner.

The following are three brief examples of how evidence challenges normal science to lead a paradigm shift.

Unobservable germs are the cause of death

Ignaz Philipp Semmelweis (1818–1865) was a Hungarian physician now known as an early pioneer of antiseptic procedures. Working in Viennese obstetric clinics, he discovered that the incidence of puerperal fever could be drastically reduced by washing hands with chlorinated lime solutions. Puerperal fever was common and often fatal in hospitals in the mid-19th century. In the Vienna General Hospital's First Obstetric Clinic, doctors' wards had three times the mortality of the midwives' wards. Despite the fact that hand washing reduced mortality to below 1%, Semmelweis's observations conflicted with the established scientific and medical opinions of the time, and his ideas were rejected by the medical community. As he could not offer any acceptable scientific explanation for his findings, some doctors, offended by the suggestion that they should wash their hands, mocked Semmelweis for it. Later, he suffered a nervous breakdown and was taken to an asylum by his colleagues. He died 14 days later after being beaten by the guards.

Semmelweis's practice only earned widespread acceptance years after his death, when Louis Pasteur confirmed the germ theory, and Joseph Lister practiced and operated using hygienic methods with great success. Germs were not observable during Semmelweis's time, so he had great difficulties showing the relationship between unobserved germs and hand washing. Still, the effects of the unobservable were seen in the death rate, which led to the great discovery of germs.

Most soldiers did not die on the battlefield but in military hospitals

Florence Nightingale (1820–1910) was an English social reformer and the founder of modern nursing. She came to prominence while serving as a manager and trainer of nurses during the Crimean War, in which she organized care for wounded soldiers. She found that wounded soldiers were poorly taken care of by overworked medical staff in the face of official indifference. Medicines were in short supply, hygiene neglected and mass infections common, many of them fatal. There was no equipment to process food for the patients.

Most of the casualties among the soldiers did not die during the battles but died after getting infected with diseases. By implementing hand washing and other hygienic practices in the war hospital, she reduced the death rate from 42% to 2%. During the Crimean War, Nightingale got the nickname 'The Lady with the Lamp in the Night' from a phrase in a report in *The Times.* After countless wars in history, an observant person during the 19th century was able to save many soldiers' lives by discovering that infections were the primary cause of their deaths and recognizing the importance of hygiene. What went through the minds of the nurses and doctors who had taken care of the soldiers in previous wars? Acceptance that suffering and deaths from all causes were inevitable, and hence normal, consequences of war?

FIGURE 4.1
The Lady with the Lamp: A painting of Florence Nightingale by Henrietta Rae, 1891.

Lack of nutrition can cause diseases

Hunger is still a common cause of death in the modern world. But what causes fatal diseases and deaths has more to do with the lack of certain nutrients than with the amount of food. For instance, scurvy is a disease resulting from a lack of vitamin C. It occurs more often in the developing world in association with malnutrition. Scurvy was described as early as the time of ancient Egypt. During the Age of Sail, it was assumed that half of the sailors would die from scurvy on a given trip. A Scottish surgeon in the Royal Navy, James Lind, is generally credited with proving that scurvy can be successfully treated with citrus fruits in 1753. However, it would be 1795 before health reformers such as Gilbert Blane persuaded the Royal Navy to routinely give lemon juice to its sailors. Why such a long delay? It took another century before the key element in the citrus fruit, vitamin C, was identified, leading to the Nobel prize for Albert Szent-Giörgyi in 1937. Before the knowledge of the essential role of vitamin C in cell metabolism, it was difficult for people to understand why so many people had to die during long sea journeys and yet the cure was indeed very simple.

Something that cannot be explained by contemporary normal science may lead to great new discoveries. Several paradigm shifts and unobservables create difficulties in the normal sciences. Indeed, unobservables in the real world are a source of ignorance in the world of minds, and ignorance has never been successfully axiomatized. Kuhn noted that after a paradigm shift, the same evidence (information) could be viewed in an entirely different way. After discovering germs and vitamins, we were able to see diseases differently. The duck-rabbit optical illusion below, made famous by Ludwig Wittgenstein, the Austrian-British philosopher, demonstrates the way in which a paradigm shift could cause one to see the same information in an entirely different way.

Part II

Probability and Inverse-Probability Inference

5

The Rise of Probability

Ignorance, incomplete information, randomness, variation and diversity exist everywhere. This means we will unavoidably face uncertainty and complexity when we make day-to-day judgements, decisions or predictions. The consequences of our actions cannot be specified or predicted with complete certainty. Probability is the most established concept to address this uncertainty. We already discussed in the previous chapters that reasoning solely based on deduction has its own limitations and entails some arbitrariness in the premises that we can agree on. Induction also cannot guarantee the truth of its conclusions. It is probability that equips induction with the necessary tool and language to deal with uncertain conclusions.

Before the middle of the 17th century, the term 'probable' – *probabilis* in Latin – meant approvable and was applicable to both opinion and action. A probable action or opinion was one that sensible people would take or hold under the circumstances. Is this approval a personal choice or a group agreement? Is it possible for all rational people to come to an agreement? This kind of definition is, in some sense, ambiguous. In the legal context, the word 'probable' can apply to events for which there is sufficient evidence. But probability has also been used to represent the degree of belief in propositions. Is the degree of belief objective or subjective? Can we all agree on uncertain propositions? Can we have an objective degree of belief in propositions independent of the human mind?

There is a general consensus on the mathematical theory of probability, a solid branch in mathematics thanks to Kolmogorov and his axioms. But we cannot say the same about its philosophical theory, which remains controversial to this day. Among statisticians, there is a trend towards neglecting the philosophy of probability. However, for serious students of statistics, we believe that the philosophy is as important as the mathematics. Firstly, it is closely related to the interpretation of probability when it is applied. Secondly, the different philosophies and interpretations have led to two distinct schools of statistics with distinct methodologies: the frequentist and Bayesian schools.

The division roughly corresponds to two facets of probability: (i) objective probability, to describe repeatable random events such as coin tosses, and (ii) epistemic probability, to describe degree of belief. When used in the first way, probability is interpreted as objective long-run frequency, while the second is considered subjective. However, philosophers consider many more theories of probability, which we shall discuss in detail later.

In its first systematic study by Fermat and Pascal, probability was associated with random events in games of chance. On the contrary, Leibniz took probability as an epistemic notion as the degree of uncertainty in the right to ownership of a piece of land. So, throughout its history, probability has represented either the chances of a certain random event happening in the real world or the epistemic degree of belief or ignorance for propositions in the world of mind.

Hacking (1975) called this the Janus-faced character of probability, Janus being the Roman god of beginnings, after whom the month of January is named. He is represented with two faces: one looking forward to the future, perhaps symbolizing stochastic laws of chance, and the other looking back to the past, perhaps symbolizing reasonable degrees of belief in propositions. The former forms an objective interpretation of probability. It refers to what happens in nature, in space and time; its statements can be checked against observed data. The latter forms an epistemic or subjective interpretation of probability. It refers to what we know in our minds and it may lack the objective testability of the former, so it instead needs a subjective commitment or a consensus agreement.

5.1 Mathematical Probability

While the probability theory had been developed since the 17th century, it existed only as a collection of mathematical results without a foundation. The axiomatization of probability – put within the axiomatization of physics – was chosen by Hilbert in 1900 as one of the outstanding unsolved problems for the new century: number 6 in his list of 23 problems. Unique features of probability, such as randomness and uncertainty, made it unclear what special axioms were needed. In the meantime, during the early 1900s, measure theory was being developed by Emile Borel, Henri Lebesgue and others, particularly after the publication of Lebesgue's thesis in 1902. Kolmogorov's contribution in 1933 was to show that no new axiom was needed to deal with probability theory. Except for the substantive interpretations, it can virtually follow the more mature measure theory.

There are alternative systems of probability axioms, but the most widely accepted system today is the one developed by the Russian mathematician Andrey Kolmogorov (1933). Let $E, F, E_1, E_2, \ldots$ stand for events, which are subsets of the sample space Ω. The events belong to a σ-algebra, a nonempty collection of subsets of S closed under complementation and countable unions. His three axioms are

Axiom 1. $\Pr(E) \geq 0$

Axiom 2. $\Pr(\Omega) = 1$

Axiom 3. (Countable additivity – CA) If E_1, E_2, ... are disjoint events, then

$$\Pr(E_1 \cup E_2 \cup \cdots) = \Pr(E_1) + \Pr(E_2) + \cdots .$$

The CA axiom is a distinguishing feature of the Kolmogorov system. When we say 'probability' without any qualifier, we refer to Kolmogorov's probability or CA probability. In Bayesian statistics and in many philosophical discussions, one instead assumes

Axiom 3'. (Finite additivity – FA) If $E_1, \ldots, E_n$ are disjoint events, then

$$\Pr(E_1 \cup \ldots \cup E_n) = \Pr(E_1) + \cdots + \Pr(E_n).$$

The corresponding theory is called the FA probability. Note, however, that although Axiom 3' refers to finite collections, the events themselves are presumed to belong to a σ-algebra. If the sample space Ω is finite, then CA and FA coincide, so the difference only matters when Ω is an infinite set. Clearly, CA implies FA, but not vice versa, so the FA axiom is weaker than CA. More differences will be discussed below.

The first two axioms trivially mean that probability is represented as a number between 0 and 1, where 0 indicates impossibility, and 1 certainty. Therefore, the third axiom is a truly defining property of probability.

In principle, anything that satisfies the three axioms can be called probability. Kolmogorov's probability is traditionally defined as a function on sets, which is natural for random events. But, by a straightforward usage of the 'and' and 'or' operations as the equivalents of 'intersection' and 'union,' one can also imagine that probability applies to propositions.

For readers familiar with the measure theory, Kolmogorov's first and third axioms look indeed like the axioms of abstract measure. The second axiom merely makes the measure finite and interpretable as the degree of uncertainty. So what makes probability unique relative to measure theory? Firstly, it is the extra notion of conditional probability, defined by

$$\Pr(E|F) = \Pr(E \cap F)/\Pr(F), \tag{5.1}$$

provided $\Pr(F) \neq 0$. The definition has the foundational power of an axiom. It appears as a definition rather than an axiom due to the abstract nature of Kolmogorov's probability. If probability and conditional probability are first defined operationally, as in subjective probability (Section 6.5), then the usage of conditional probability can be introduced via an axiom of product rule:

$$\Pr(E \cap F) \equiv \Pr(E|F)\Pr(F).$$

Philosophers view this as more elegant than the alternative way to use the definition (5.1). By adding this axiom, the probabilities of $E \cup F$ and $E \cap F$ are covered by two axioms instead of one axiom and one definition.

Secondly, we have the notion of randomness, which is simply captured by the independence of two events. E and F are said to be independent if

$$\Pr(E|F) = \Pr(E).$$

Alternatively, $\Pr(E \cap F) = \Pr(E)\Pr(F)$ if and only if E and F are independent. All of Kolmogorov's celebrated theorems in probability are based on the sequence of independent random variables. Moreover, theoretical analyses of dependent random variables must rely on the construction based on independent ones.

Boy or Girl Paradox

A conditional probability problem known as the Boy or Girl Paradox, or The Two Child Problem, appeared in an article by Martin Gardner in *Scientific American* in 1958. The paradox highlights the delicate effects of information and precise formulation in probability and statistical problems. It is easy to come up with questions that are ambiguous and lead to conflicting answers. The failure to be precise is the main issue with, for instance, the paradox of the ravens (Chapter 18).

Mr. Smith has two children and one of them is a boy. What is the probability that the other is a girl? If you quickly guess 1/2, you're not alone. Conditional probability is conceptually easy enough to understand if both E and F are observable events. However, probability is often non-intuitive, and extreme care is needed in formulating how information is gathered and presented to us. Suppose *we randomly sample one family from a collection of families with two children* – where the father happens to be Mr. Smith – and observe that the *older* child is a boy. What is the (conditional) probability that the other child is a girl? Assuming the gender of each child is independent of each other, knowing that the first born is a boy does not tell us anything about the second born, so the answer is 1/2, which matches our intuition.

Instead, in the Paradox, we are only told that one of the children is a boy, but not about the birth order. The information seems clear enough, but there is actually an ambiguity: Does it mean that *exactly one* is a boy? Or *at least one* is a boy? If the former, then the probability is 1; if the latter, then 2/3. To see this, with obvious abbreviations, the sample space of families contains labels BB, BG, GB and GG, each with probability 1/4. The former is

$$\frac{\Pr(BG, GB)}{\Pr(BG, GB)} = 1,$$

while the latter is

$$\frac{\Pr(BG, GB)}{\Pr(BB, BG, GB)} = 2/3.$$

So, neither is 1/2!

Next, suppose that, instead of *taking a random family,* we draw a random child from *the collection of children* from two-child families and observe a boy – whose father happens to be Mr. Smith. What is the probability that the other child in the boy's family is a girl? Intuitively, the child's gender does not give information about the gender of his sibling. But at this point, you may not trust your intuition. So, imagine that we sample a child randomly, then ask him/her the gender of his/her sibling. Being explicit about the order of observations (not the age order, which we do not know), the sample space contains B_1B_2, B_1G_2, G_1B_2 and G_1G_2, each with a probability of 1/4. So, the conditional probability of a girl sibling after seeing a boy is

$$\frac{\Pr(B_1G_2)}{\Pr(B_1B_2, B_1G_2)} = 1/2,$$

now matches our original intuition again. If this explanation is not satisfactory for you, imagine 100 families of two children, so there are 25 families each with BB, BG, GB and GG children. In total, there are 100 boys and 100 girls, and of the 100 boys, 50 are from the BB families and 50 from mixed families (BG or GB). So if the chosen child is a boy, there is a 50-50 chance that he comes from a mixed family, in which case the other child is a girl.

Finally, instead of making explicit statements about sampling, suppose we just observe Mr. Smith walking down the road with his boy. Assuming that he has two children, what is the probability that the other child is a girl? This anecdotal version has no answer, since it is ambiguous which sampling model applies: Is it based on families – as represented by Mr. Smith? Or based on children – as represented by the boy? We have shown above that these lead to two different answers. Also different would be if *it is Mr. Smith himself* who tells us that he has one boy. Assuming he has two children and no reason to lie or mislead us, it would be absurd for him to say 'I have one boy' if in fact he has two. Therefore, in this scenario, the other child must be a girl.

5.2 Countable vs Finite Additivity

For some people, Kolmogorov's CA axiom – involving an infinite number of sets – is not self-evident. CA is often justified by its success in leading the theory to landmark theorems. However, many philosophers and prominent Bayesian statisticians, such as de Finetti and Savage, claim that it is sufficient and more self-evident to have FA as the third axiom. Savage wrote that he knew 'of no argument leading to the requirement of countable additivity... therefore seems better not to assume countable additivity outright as a postulate ... .' Kolmogorov (1933) himself argued that the countable additivity axiom was essential theoretically, but 'it is almost impossible to elucidate its empirical meaning.'

Since CA implies FA, the latter imposes a weaker constraint. This means that all models that satisfy CA automatically satisfy FA, but there are probability models allowed under FA, but not under CA. Thus, FA appears to allow for a larger class of probability models. There is, however, a serious theoretical price for that. For example, under FA we may have non-conglomerability (Scherwish et al., 1984). Let $\mathcal{B} \equiv \cup_i B_i$ be a countably infinite partition of the sample space S. To be non-conglomerable with respect to $\mathcal{B}$, there exists some event E where

$$\alpha_1 \leq \Pr(E|B_i) \leq \alpha_2, \text{ for all } B_i \in \mathcal{B},$$

but

$$\Pr(E) < \alpha_1 \text{ or } \Pr(E) > \alpha_2.$$

Specifically, it may happen that $\Pr(E) > \Pr(E|B_i)$ for all B_is. This is a paradoxical property. To see it, let's imagine two bets/lotteries: (i) a simple one involving just E vs not-E, and (ii) a complex one, still involving E vs not-E but with different tickets offered according to which B_i occurs first. Because B_is form a partition, the first lottery can be thought of as a version of the second lottery where the price is fixed at $\Pr(E)$. The result $\Pr(E) > \Pr(E|B_i)$, for all B_is, means that we have a lottery with the same tickets as another lottery, but each ticket is more likely to win. This violates common sense, which is encapsulated in the so-called comparative principle (Easwaran, 2013).

In contrast, this cannot happen under CA probability, since here we have conglomerability

$$\Pr(E) = \sum_i \Pr(E|B_i)\Pr(B_i),$$

so we have $\alpha_1 \leq \Pr(E) \leq \alpha_2$. This is as expected intuitively: as the (weighted) average, $\Pr(E)$ must respect the same lower and upper limits as $\Pr(E|B_i)$'s. This means that the basic mathematical results of the CA probability may not hold for the FA probability without further justifications.

An example that seems to convince both de Finetti and Savage about the merits of FA over CA is the so-called de Finetti lottery or God's lottery, which is a fair lottery on the set of positive integers $\mathcal{N}$. De Finetti's reasoning (1970, vol. 1, p.122) was not so clear: Countable additivity 'forces me to choose some finite subset of them ... to which I attribute a total probability of at least 99%... .' Agreeing with de Finetti,

Savage (1972, p. 43) wrote 'many of us have a strong intuitive tendency to regard as natural probability problems about the necessarily only finitely additive uniform probability densities on the integers, on the line and on the plane.'

We can in principle play God's lottery: I write down any positive integer and you guess which number or what set the number belongs to. To be fair, the winning probability must be distributed uniformly among the infinite number of integers. Suppose $\Pr(X = k) = c \geq 0$. For any value $c > 0$, under CA, $\Pr(\mathcal{N}) = \Pr(\bigcup_k [X = k]) = \sum_k \Pr(X = k) = \infty$, which violates the second axiom. So c must be zero; but now $\Pr(\bigcup_k [X = k]) = 0$, again violating the second axiom. So, God's lottery is not allowed under CA. But it is *allowed* under FA: the probability of any finite collection of integers is zero, so there is no contradiction with $\Pr(\mathcal{N}) = 1$.

Even though it is allowed under FA, a uniform distribution on the positive integers has many counterintuitive properties. For example, since each integer has zero probability, which values carry the probability mass? Under CA, defining the probability for each singleton integer, say $\Pr(X = k) = 2^{-k}$, for $k = 1, \ldots$, determines the probability of any event. But, under FA, by defining $\Pr(X = k) = 0$ for any $k > 0$, we only know that the probability of any finite set is zero and that of any co-finite set is one. But what about, for example, the probability of infinite sets such as even integers? or odd integers? or multiples of three? As far as we can search in the literature, we have never seen a clear and rigorous answer to these questions.

There have been many attempts on both the philosophical and mathematical sides to make sense of FA. In an amusing article, McCall and Armstrong (1989) listed five scenarios that highlight the counterintuitive nature of God's lottery. Which would you choose between the two options in each of the following:

1. (A) All tickets numbered 1 to 100 or (B) all tickets numbered 101 to 1,000,000? Theoretically, A and B have the same probability of zero, but intuitively, B has a higher probability.

2. (A) All even-numbered tickets or (B) all multiples of three? Intuitively, A is more numerous, but Cantor already told us that these two sets have the same cardinality. That is, there is a one-to-one map between A and B, so by the fairness of the lottery they should have

the same probability. Unless 'fairness' is to mean something else in God's lottery?

3. (A) Multiples of 3 or (B) multiples of 47198? The same reasoning as in the previous choice applies here but somewhat more extreme. It's hard to resist A over B.

4. (A) Multiples of 47198 or (B) all numbers less than $10^{10^{10}}$? Theoretically, B has probability zero. Although we do not know what the probability of A is, the set feels very sparse.

5. (A) Multiples of six, or (B) multiples of three? This is interesting. If the winning ticket is in A, then it must be in B. However, if the winning ticket is in B, it is not necessary in A. It is absurd not to choose B. But, again, these two sets have the same cardinality.

Perhaps de Finetti and Savage were just being consistent with their choices of axioms, which imply FA probability. De Finetti defined probability as a coherent betting price (Section 6.5), which is the price that protects one from the Dutch Book attack. So what is the proper price to play God's lottery, say, to bet on a finite set? Theoretically, it must be zero or at most infinitely small. And what is the proper winning prize? Theoretically, to be fair, it should be infinitely large. So, we have a counterintuitive and counter-empirical situation where it is irrational to pay anything for a lottery that promises an infinite prize. This points to a clear remedy: Williamson (1999) showed that a requirement that the bets and prizes be finite leads to a subjective probability that satisfies CA.

Similarly, Savage started with seven axioms of rational preferences that imply FA probability (Chapter 7). There were later modifications of those axioms – for instance, Villegas's (1963) monotonicity – that lead to CA rather than FA.

5.3 What Is Probability?

The probability in Kolmogorov's axioms is an undefined entity. It is comparable to the concept of 'point' in abstract geometry, which is not necessarily connected with the point we understand in daily life;

for instance, a function can be a point in a Hilbert space of functions. The intuitive meaning of the concept can come when some of the theorems are connected to observations. For example, given a sequence of Bernoulli trials, the law of large numbers theorem states that the sample proportion converges to the true probability. This mathematical result naturally suggests a frequentist long-run interpretation of probability. If you believe that this is the only valid interpretation of probability, then you will be classified as a frequentist. If you don't agree, what are your options?

There are two broad interpretations of probability: epistemic and objective. Due to the flexible meaning of the word 'objective,' these words are not completely satisfactory. Your 'objective opinion' might mean impartial or unprejudiced, while 'objective reality' means verifiable physical reality. Even the seemingly obvious term 'objective existence' is ambiguous, because it might refer to objects with real physical existence, such as tables and chairs, or to entities that only exist by consensus, such as money, prices, customs and laws, corporations, nation states, etc. The historian Y.N. Harari (2015) calls the latter 'imagined reality,' the fruit of a uniquely powerful human imagination. This is why we later interpret probability in more than two ways. We first illustrate how different people have different views depending on their area of expertise.

Epistemic interpretation roughly relates to the state of mind or degree of belief in the world of mind, whereas objective interpretation has to do with human-independent entities in the objective material world. Ramsey (1931) wrote,

> the theory of probability is taken as a brand of logic, the logic of partial belief and inconclusive argument; but there is no intention of implying that this is the only or even the most important aspect of the subject. Probability is of fundamental importance not only in logic for the world of mind, but also in statistical and physical science in the real world, and we cannot be sure beforehand that the most useful interpretation of it in logic will be appropriate in physics as well. Indeed, the general difference of opinion between statisticians who for the most part adopt the objective (frequentist) theory of probability for events and logicians who mostly reject it, renders it likely that the two schools are really discussing different things. The only difference is that the word "probability" is used by logicians in one sense for the world of mind and by statisticians in another sense for the real world. The conclusions we shall come to as to the meaning of probability in logic must not, therefore, be taken as prejudging its meaning in physics.

The philosopher Gillies (2000) argued that the epistemic notion of probability is appropriate for the social sciences, whereas the objective notion is appropriate for the natural sciences. His objective probability is associated with repeatable conditions for independent outcomes. Independent repeatable outcomes are prevalent in natural sciences but rarely observed in social sciences in general, including economics and finance.

The Nobel laureate in economics John Hicks (1980) noted,

> I have myself come to the view that the frequentist theory, though it is thoroughly at home in many of the natural sciences, is not wide enough for economics. ... At the end of 1944... the probability that the European war would come to an end within a year. This was a probability which, at that date, most people would have assessed to be a high one. But it is quite clear that it does not fall within the frequentist definition: it is not a matter of trials that could be repeated. We cannot avoid this kind of probability in economics. Investments are made, securities are bought and sold, on the judgment of probabilities. ... Probability, in economics, must mean something wider.

The famed billionaire investor Soros (1987) wrote,

> the event studied by social sciences have thinking participants: natural phenomena do not. The participants' thinking creates problems that have no counterpart in natural sciences. ... in quantum physics, it is only the act of observation which interferes with the subject matter, not the theory of uncertainty, whereas in the case of thinking participants, their own thoughts form part of the subject matter to which they relate...

Last but not least, according to the physicist Heisenberg,

> probability represents a mixture of two things, partly a fact and partly our knowledge of a fact. ... Given the initial situation, the electron moving with the observed velocity at the observed position; 'observed' means observed within the accuracy of the experiment. It represents our knowledge ... The error in the experiment does – at least to some extent – not represent a property of the electron but a deficiency in our knowledge of the electron. Also, this deficiency of knowledge is expressed in the probability function. Even in classical physics, due to the observer effect, it would be natural to consider a probability for the initial values of the coordinates and velocities and therefore something very similar to the probability function in quantum mechanics.

Classical statistical mechanics studies phenomena caused by the mass behaviour of particles. Apart from occasional collisions, the particles move independently of each other. The relationships themselves are deterministic, but the consequences of their mass behaviour can only be explained by probability. Here, probability proves to be useful in describing mass behaviour. Compared with quantum mechanics, Heisenberg noted,

> Only the necessary uncertainty due to the uncertainty relations is lacking in classical physics. ... one can, from the laws of quantum theory, calculate the probability at any later time and can thereby determine the probability for a measurement giving a specified value of the measured quantity. We can, for instance, predict the probability of finding the electron at a later time at a given point in the cloud chamber. It should be emphasized, however, that the probability function does not in itself represent a course of events in the course of time. It represents a tendency for events and our knowledge of events. The probability can be connected with reality only if one essential condition is fulfilled: if a new measurement is made to determine a certain property of the system. Only then does the probability allow us to calculate the probable result of the new measurement. ... we change over again from the "possible" to the "actual."

So, in quantum mechanics, the observer's actions interact with particles, which is somewhat close to what Soros stated.

5.4 Pascal's Wager

In Section 2.2 we discussed the deductive proofs of God's existence. What if you can't accept the axioms? Are you then an atheist? Let us define G as the proposition that God exists. Kant said that we cannot determine its truth value by empirical or rational proofs (although he did not consider himself an atheist as he did not deny the possibility of the existence of God). Pascal's contribution is to show that we do not need to know the truth of G to decide what to do.

Pascal's wager marked the first use of probability for decision making under uncertainty: 'God is, or He is not. But to which side shall we incline?' Being a gambler, Pascal imagined a game being played where heads or tails would turn up, representing God's existence or absence with a certain probability. You are in this unavoidable game until death, after which point you will be able to observe the outcome. What will you wager? Even if one is ignorant of the truth, Pascal argued that a rational person should live properly as though God existed. If God does not exist, such a person will have, at most, only a finite loss (some pleasures, luxury, etc.), while standing to receive infinite gains (as represented by eternity in Heaven) and avoid infinite losses (eternity in Hell).

Pascal proposed to represent the truth of G by a probability:

$$\Pr(G) = \Pr(G \text{ is true}) = \Pr(I(G) = 1) = p, \quad 0 \leq p \leq 1.$$

According to his statement, his probability $\Pr(G)$ can be interpreted as epistemic or objective probability. He may not know these two modern interpretations, but it does not matter because it is difficult to compute his probability. He conjectured that his probability p is like the long-run frequency of heads in coin tossing, but unknown. It is as though we tossed the coin but did not yet see whether the outcome was heads or tails. This coin will not be tossed again in our lifetime. But, we can still act rationally. Because $\Pr(G)$ is like a frequency in coin tossing, Pascal conjectured that we can compute the expected return for rational decision making. Pascal's wager can be described as a decision-making exercise with the payoff values given in Table 5.1. You have two possible actions. One is to believe in the existence of God, and the other is not to believe.

TABLE 5.1

Pascal's wager: the rewards assumed if one chooses to believe or to disbelieve in the existence of God.

	God exists (G)	God does not exist (not G)
Believe (B)	$+\infty$(infinite gain)	-1 (finite loss)
Disbelieve (not B)	$-\infty$(infinite loss)	$+1$ (finite gain)

If $\Pr(G) = 0$, then $E(B) = -1 < E(\text{not } B) = 1$, whereas if $\Pr(G) > 0$ the option of living as though God exists (B) has the expected reward

$$E(B) = \infty \times \Pr(G) - \Pr(\text{not } G) = \infty,$$

whereas the option of living as if God does not exist (not B) has the expected reward

$$E(\text{not } B) = -\infty \times \Pr(G) + \Pr(\text{not } G) = -\infty.$$

Thus, to get the infinite reward, a person should live as if God exists as long as they entertain a non-zero probability for it.

If Pascal's probability $\Pr(G)$ is an epistemic personal probability, then $E(B)$ becomes the expected return for the bet. However, according to economic theories, real people tend to dismiss infinite utilities as unrealistic. For example, if people complain that death is so terrible, it should be assigned a utility of $-\infty$. But if so, nobody should do anything whatsoever that involves the slightest probability of death. No one should ever leave home because a meteor might fall on their head. In fact, judging from their driving behaviour, some people don't seem to fear their deaths that much. And real people will not appreciate the infinite reward either. For example, a trillion dollars should be more than enough to make one happy beyond their imagination. Moreover, unfortunately, Pascal's infinite reward or punishment come in the afterlife, so we must apply severe time discounting. So, consider instead the finite rewards in Table 5.2.

TABLE 5.2
Pascal's wager with finite rewards. The value of A varies between individuals according their reward perception.

	God exists (G)	God does not exist (not G)
Believe (B)	+100	−1
Disbelieve (not B)	−100	A

Under finite rewards, the expected values are

$$\begin{aligned} E(B) &= 100\Pr(G) - \Pr(\text{not } G) = 101\Pr(G) - 1 \\ E(\text{not } B) &= -100\Pr(G) + A(1 - \Pr(G)) = A - (100 + A)\Pr(G). \end{aligned}$$

Choosing to believe in God is consistent with $E(B) > E(\text{not } B)$, or

$$\Pr(G) > (A + 1)/(201 + A).$$

To illustrate, first note that A is the reward of living the life as if God does not exist when in fact He does not. Suppose you choose to believe in God even if you're offered a reward of $A = 10$; it follows that your belief is $\Pr(G) > 11/211 \approx 0.052$. Alternatively, we can interpret the result this way: for someone with $\Pr(G) = 0.05$, a reward larger than $A = 10$ can turn them into a non-believer. Not surprisingly, your belief must be stronger if you're not persuaded by a larger reward. For $A = 100$, comparable to the reward given to the believer when God truly exists, $\Pr(G) > 0.33$. In the extreme, for someone with $\Pr(G) = 1$, no amount of reward can persuade them to change their mind. Overall, this example illustrates that an observed behaviour or decision is connected to and can reveal subjective probability and reward perception.

6

Philosophical Theories of Probability

Statisticians typically recognize only two interpretations of probability: Bayesian/subjective and frequentist. However, when Fisher was defending his fiducial probability, he referred to the probability of the 'early writers,' which was presumably neither Bayesian nor frequentist, because he was critical of both. In cases where fiducial probability is conceptually acceptable, how should we interpret it?

The philosophical issue is indeed interpretational, not mathematical; in statistical theory and modelling, there is a virtual consensus around Kolmogorov's probability. Mathematically, Kolmogorov's axiomatic foundation puts probability as the legitimate child of the more mature measure theory. The celebrated laws of large numbers in Kolmogorov's theory naturally give probability a long-run frequency interpretation. But people do bet on specific events, which can only mean that they also have subjective probability in mind.

We follow Gillies (2000) in his categorization of six distinct philosophical theories: (i) classical, (ii) logical, (iii) subjective, (iv) frequency, (v) propensity and (vi) inter-subjective. The logical and subjective interpretations can be considered epistemic, the frequency and propensity ones objective. The last theory is a hybrid theory, actually the one we consider most compelling for the objective probability of single events. However, the term 'inter-subjective' sounds like a contradiction and a variation of the subjective theory. For the reasons given in Section 6.6, we shall call it 'consensus theory.' There is a disagreement among philosophers about whether the classical theory is epistemic or dual epistemic-objective; in either case, the classical theorists never made any explicit distinction between the two main interpretations.

The frequency theory is, of course, closely connected to the frequentist school in statistics, but the connection is not as close as the Bayesian school to the subjective theory. Our guess is that all Bayesian statisticians subscribe to the subjective theory, but, as we explain later, frequentist statisticians actually do not follow the axioms of frequency theory at all. The probability view of frequentists, as expressed, for example in the orthodox teaching of introductory statistics, is somewhat a mixture of the frequency and propensity theories. Unless stated otherwise, we use the term 'frequentist' to refer to the frequentist school of statistics or its adherents. The philosophical theory is referred to as frequency theory, not frequentist theory.

6.1 Classical Theory

The first era of classical probability – involving the 'early writers' such as Fermat, Pascal, Galileo, Huygens, Jakob Bernoulli and De Moivre in the 17th and 18th centuries – grew primarily out of the games of chance. Among the landmark publications of this era, we find Jakob Bernoulli's *Ars Conjectandi*, published in 1713; here he proved his 'Golden Theorem' that the sample proportion converges in probability to the true binomial proportion. De Moivre's *The Doctrine of Chances* first appeared in 1718; its second edition (1738) contained the proof of the normal approximation to the binomial distribution, a precursor of the central limit theorem later proved for more general random variables by Laplace in his 1812 book. Bayes's *Essay Towards Solving a Problem in the Doctrine of Chances* (1763) and Laplace's *Memoir on the Probability of Causes by Events* (1774) were among the most influential papers in statistics (Stigler, 1986), though Bayes's paper was mostly unnoticed until early in the 20th century.

In this period, probability was defined operationally as the proportion of favourable outcomes among equally probable cases. This seems natural given its application in the game of chance involving coins, dice or cards. The definition is not problematic for finite sample spaces. It is also obvious that the resulting probability satisfies the finitely additive probability axioms (Section 5.1). In terms of later theories, the simple operational definition plus its primary application in gambling actually give probability dual interpretations as (i) a subjective quantity – which allows one to bet rationally on single events, or (ii) a long-term frequency as given by Bernoulli's Golden Theorem. Perhaps this is what Fisher had in mind, that those 'early writers' did not separate the epistemic and objective meanings of probability; see also Daston (1988, p. 191).

However, some philosophers such as Gillies (2000) believed that classical probabilists were more on the epistemic side, particularly due to the influence of Laplace. The classical theory culminated in Laplace's magnum opus *Analytical Theory of Probability*, published in 1812, which became a major reference for the rest of the century. As a major figure in the Enlightenment era, Laplace believed in universal determinism, a principle that had we all the information and intelligence needed, nature was completely predictable. He pointed to the success of Newton's theory of gravitation in explaining the planetary movements. As he wrote in 1814 at the beginning of *A Philosophical Essay of Probabilities*:

> ALL events, even those which on account of their insignificance do not seem to follow the great laws of nature, are a result of it just as necessarily as the revolutions of the sun. ...
>
> ...
>
> Present events are connected with preceding ones by a tie based upon the evident principle that a thing cannot occur without a cause which produces it. This axiom, known by the name of *the principle of sufficient reason*,...

Therefore, probability is an expression of human ignorance and partial information, not an inherent/objective state of nature. Here is the idea. Suppose you toss a coin; if you have *all* the proper measurements and background theory, then you should be able to predict exactly whether it's going to land heads or tails. There is no probability involved. But, due to our ignorance, we can only say that there is a probability of 0.50 that we will get heads. In this sense, probability is epistemic, as it is a measure of human knowledge, not an objective quantity. Laplace attributed differences of opinion about the same thing to being due to different degrees of information as well as 'the manner in which the influence of known data is determined.' The latter points to the important role of probability theory for applications in the 'moral sciences,' such as testimonies, the decisions of the Assembly, the judgements of tribunals, tables of mortality, etc.

As with previous probabilists, Laplace relied on the principle of insufficient reason, first formulated by Jakob Bernoulli, to justify the equal probability assignment to *events of the same kind* for which we are otherwise ignorant. This perspective will affect how we model probability.

Suppose you're told that a coin is biased, but the bias can be either positive or negative. If you're an objectivist, you believe there is an underlying probability $\theta \neq 0.5$. But if you believe in the principle of insufficient reason, you still have equal ignorance of heads and tails, so would then assign probability 0.5 to each.

Bertrand's and von Mises's paradoxes

When used unthinkingly, the principle of insufficient reason can lead to paradoxes, the famous ones being Bertrand's random-chord paradox (1889) and von Mises's wine-water paradox. The principle feels natural in discrete cases, but a serious issue arises when the sample space is continuous. In this case, the principle implies that the uniform distribution represents complete ignorance. But there could be natural alternatives with the random quantities that can be made uniform, and each choice leads to a distinct answer.

Bertrand's paradox starts with a seemingly innocuous question: Consider the circumcircle (outer-circle) of an equilateral triangle; a chord of the circle – a straight line whose endpoints lie on the circular arc – is chosen at random. What is the probability that the chord length is larger than the side of the triangle? The problem is, how do we choose such a chord at random? There are three distinct – and equally convincing – ways of doing that, each with its own uniform distribution:

(i) Choose two points at random on the circular arc, then construct a chord by connecting the points.

(ii) Choose a radius of the circle at random, then choose a point on the radius at random. Finally, draw a chord perpendicular to the radius.

(iii) Choose a random point inside the circle, then draw a chord at a random angle.

And, unfortunately but not surprisingly, these three methods lead to different answers, giving probabilities equal to $1/3, 1/2$ and $1/4$, respectively. It's unfortunate, because the paradox calls into question our intuition about what 'random' means. Jaynes (1973) argued that we can get the second solution as a unique solution by imposing shift and scale invariance conditions. However, as shown in Drory (2015), Bertrand's two other solutions can also satisfy some shift, scale and rotation invariance conditions.

In von Mises's paradox, some amounts of wine and water are mixed so that the amount of one liquid is less than three times that of the other. That is, if R is the ratio of wine to water, then R is between 1/3 and 3. What is the probability that the amount of wine is less than twice the amount of water? We are looking for the probability that

$$R = \text{wine/water} \leq 2.$$

The principle of insufficient reason implies the uniform distribution for R on the interval $[1/3, 3]$. Thus,

$$\Pr(R = \text{wine/water} \leq 2) = (2 - 1/3)/(3 - 1/3) = 5/8.$$

But we could equally ask for the probability that

$$W = 1/R = \text{water/wine} \geq 1/2.$$

Our ignorance again leads to the uniform distribution of W on $[1/3, 3]$, but now we get

$$\Pr(W = \text{water/wine} \geq 1/2) = (3 - 1/2)/(3 - 1/3) = 15/16.$$

The uniform distribution for the R scale and the $1/R$ scale leads to very different answers. The use of a flat probability distribution means that it contains information about the scale, of which we are supposed to be ignorant.

Book and coin paradoxes

Simpler paradoxes of the principle of insufficient reason can be constructed: Suppose you pick a book at random from a library. You have no idea if the cover is green or not green. According to the principle, each should have a probability of 0.5. But you can also ask whether the cover is red or not red. Should you now assign the probability of 0.5 to red? If you say yes, then there is an immediate problem, since the total probability is greater than one. This kind of paradox can be avoided by careful enumeration of book covers into indivisible colours of the same kind, though we can still see problems arising. But Bertrand's and von Misses's paradoxes are much more resistant to easy resolution, because they are due to an inherent lack of precision in human language.

For the coin paradox, suppose that a device tosses a coin twice and you have no idea of its potential biases. Treating the outcomes HH, HT, TH and TT as equally likely, then, for instance, $\Pr(HH) = 1/4$. But what if you care only about the number of heads? Then the outcomes 0, 1 and 2 should be equally likely, so $\Pr(2) = 1/3$. This is of course paradoxical, since these are probabilities of the same outcome! You might immediately dismiss the latter as nonsensical, but in Section 9.1 we show that it is not unreasonable. In fact, it corresponds to the standard Bayesian calculation.

6.2 Logical Theory

In the Cambridge circle, under the influence of philosophers such as W.E. Johnson, G.E. Moore and Bertrand Russell, with his *Treatise on Probability*, Keynes (1921) proposed a logical theory of probability as a degree of *rational belief*. It is also meant to apply to propositions rather than just events. Keynes was actually motivated by the problem of ethics, where one cannot be sure of the consequences of one's actions but can only assess their probable effects. 'Rational' is to mean independent of subjective preferences, such that rational intelligent individuals can agree as they do in classical logic. Logical probability is meant to be a 'degree of partial entailment' in the following sense. In standard mathematical logic, one says 'A implies B' to mean that if A is true, then B is 100% guaranteed to be true. Thus, in logical probability, if A is true then B is only true up to a certain probability, not 100% guaranteed. That is, B partially follows from A. Therefore, probability expresses a logical relationship between A and B and, as such, it is an extension of classical deductive logic.

We can see the immediate relevance of logical probability in inductive inference. Let A be a collection of evidence – say we have observed 10 black ravens – and B a general hypothesis that all ravens are black. We face the inductive problem: We know B is not fully guaranteed by A, but can we logically ask what the probability of B given A is? This is exactly the same question that Bayes famously tried to answer in 1763 (Chapter 8), which is how to learn from data. Bayes's idea grew into a whole school of Bayesianism, though it comes in subjective and objective flavors. To some extent, the probability in objective Bayesianism is closest in spirit to Keynes's logical probability; see more later in Section 6.6.

What are specific examples of logical probability? In Keynes's conception, probabilities are only partially ordered, and not all probabilities have numerical values (or, perhaps, all propositions have a probability of being true, but we are not always able to 'exercise practical judgement' to know what they are). Some probabilities are only qualitative. And, there is a complex relationship between probabilities: Some are comparable, so they can be ordered by magnitude, even though they might only be known qualitatively. For example, the probability of a war between Russia and NATO within the next 10 years is greater than one between Russia and China. But some probabilities are not even comparable; for example, the probability of detecting extraterrestrial life in the universe and the probability of World War III within the next 50 years.

Perhaps Keynes was simply being realistic about our limited practical ability to assess the probability of an arbitrary proposition. But this invites one major criticism: The theory is simply too vague to provide a mathematical basis for probability. It seems safe to say that Keynes's theory never had any direct impact on statistics or mathematics. Indirectly, however, it may have influenced Jeffrey's objective Bayesian probability (Jeffreys, 1961). Also, it may have led the young Frank Ramsey to come up with his theory of subjective probability; see below.

In Keynes's construction, the only case where we can get numerical probabilities is when we can apply the principle of indifference – the good old principle of insufficient reason – for equiprobable alternatives. For example, let A be the setup for a fair throw of a coin. Then we are indifferent to the outcomes, heads or tails, so the logical probability of each is 0.5. Keynes's contribution is then to recognize this probability as an objective quantity agreed to by rational minds. However, the indifference principle brings out the second major criticism of the theory: As we have described above, the principle leads to Bertrand's and von Mises's paradoxes.

Other logical theories

One distinguishing characteristic of logical probability is that it can apply to propositions rather than just events. For propositions, we use the operators 'and' and 'or' instead of intersection and union. One weakness of Keynes's idea is that it was not built axiomatically. Jaynes (2003) presented another version of logical probability based on three 'desiderata,' sort of informal axioms, that are strong enough to lead to formal probability axioms. If we disagree with the construction, we would have to specify which desiderata we do not like, which is what we have to do if we disagree with an axiomatic system. Here they are:

(a) Representation of degrees of plausibility by real numbers
(b) Qualitative correspondence with common sense
(c) Consistency

The consistency requirement may sound vague and arbitrary, but it means classical logic for true-false propositions, and basic arithmetic and calculus for numerical and functional analysis. Jaynes showed that these desiderata imply finitely additive probability on the space of propositions. To obtain numerical probability in specific cases, such as a coin toss or a throw of the die, etc., Jaynes relied primarily on the invariance principle, which is a physical version of the principle of insufficient reason. He claimed, for example, that this solves Bertrand's random chord paradox, although this is disputable (Drory, 2015).

The philosopher Rudolf Carnap (1891–1970), a major figure in the Vienna circle, spent the last three decades of his life working on the logical theory of probability. He called it 'probability$_1$' to represent epistemic probability as a degree of confirmation for propositions, as opposed to 'probability$_2$' for statistical or long-term frequency. Like Keynes, he also viewed logical probability for induction as a continuous extension of the classical logic for deduction. His views, summarized in his book *Logical Foundations of Probability*, can perhaps be called objective Bayesian.

Late in his life, Fisher wrote a series of papers on the nature of probability. Although he did not develop any detailed theory, it is clear that he had logical probability in mind. In *The Nature of Probability* (1958), he wrote down explicitly a set of 'requirements' for probability to be 'a well-specified state of logical uncertainty':

(a) There is a measurable reference set (a well-defined set, perhaps of propositions, perhaps of events).

(b) The subject (that is, the subject of a statement of probability) belongs to the set.

(c) No relevant subset can be recognized.

The first two requirements are rather obvious, though note the inclusion of propositions as the subject of probability. The third requirement, which he considered 'the most novel,' is closely connected to the concept of reference class explained in Section 6.4. We shall also give more theoretical details in Chapter 13.

6.3 Frequency Theory

Due to Laplace's influence, during the 19th century, probability was seen epistemically as a measure of partial ignorance, which can also reflect a degree of belief. A systematic formulation of an objective frequency theory emerged in the mid-19th century, particularly in *The Logic of Chance* by John Venn in 1866. Statements of probability are to be interpreted as a prediction of the long-run frequency of repeated events. This is of course supported by a well-recognized empirical result that the sample proportion becomes more and more stable as the sample size gets larger.

Early in the 20th century, the theory was further developed in the Vienna circle, particularly by Hans Reichenbach and Richard von Mises. In his 1928 book *Probability, Statistics and Truth*, von Mises likened probability theory to other areas of science, such as mechanics or physics, where concepts like force or work have measurable quantities. His theory depends heavily on the concept 'collective' $C \equiv (c_1, c_2, \ldots)$, an infinite sequence of experimental outcomes c_is from the sample space. For instance, $HHTTHTHTHHTH...$ is one collective of coin tosses. Given n tosses, the sample proportion of heads is the proportion of Hs in the n tosses, which is $n(H)/n$. For an arbitrary collective C, we can then state the first axiom of the theory:

Axiom 1: The limit $n(A)/n$ as $n \to \infty$ exists, where $n(A)$ is the number of instances in C where event A occurs.

The probability $\Pr(A|C)$ is then defined as the limit. This is an operational definition of probability, though one might argue whether the limit to infinity is really operational. Von Mises strongly rejected the use of probability for single events, such as a war between Russia and NATO by 2030. This is a familiar point of view often expressed by frequentist statisticians.

In Kolmogorov's theory, von Mises's Axiom 1 appears essentially as a theorem: the *strong* law of large numbers (SLLN). It is only essentially, because strictly speaking, the SLLN does not guarantee convergence for all sequences and some technical conditions are needed. The axiom, however, specifies that the limit exists for any collective. The convergence in Axiom 1 cannot be the *weak* law of large numbers (WLLN), a theoretically much simpler version of LLN that specifies only a convergence in probability, because it would have a circularity problem.

The SLLN theorem clearly specifies the conditions that are needed for the convergence to hold. There are several versions of SLLN; the standard SLLN requires an independent and identical (iid) sequence of random variables with finite expected value. The requirements of identical distribution and finite expectation are conceptually relatively easy to satisfy, but independence is harder. In Kolmogorov's theory, independence follows from the definition of conditional probability, which is, of course, not possible in von Mises's scheme, as it would face a circularity problem. Independence, however, is not the only sufficient condition for the SLLN. In the theory of stochastic processes, sequences satisfying the convergence in Axiom 1 are called *ergodic*. Again, there are conditions – generally called mixing conditions – that can guarantee ergodicity. The iid condition is just one of the simplest mixing conditions. How is this condition specified in frequency theory?

For this purpose, von Mises introduced an axiom of randomness based on the empirical result in gambling:

Axiom 2: One cannot beat a gambling system by making a series of decisions on when to gamble and when not to gamble.

This is equivalent to postulating that there exists a perfect unbeatable gambling system, which must depend on randomness. So, randomness is embodied in a perfect gambling system, but what is it mathematically? In the gambling setup, the collective $C \equiv (c_1, c_2, \ldots)$ represents the serial outcomes of a random gambling system, and you gamble on each outcome. If A is the event you gamble on, you win in step i if A occurs: i.e., $c_i \in A$. It is useful to think intuitively of the collective as a realized sequence of iid random variables. But since the iid concept is not available in von Mises's system, its properties must somehow be captured by a different concept. Thus, in Axiom 2, von Mises introduced selection: a sequence of 1–0 binary decisions, representing when you choose to gamble or not gamble. For example, you may decide to gamble every 5th step, or every 17th step, or at every step given by the Fibonacci series, etc. In principle, the axiom implies that any sub-collective of C should produce the same probability as $\Pr(A)$ derived from the entire collective C.

But what mathematical rule is there to stop you from choosing to gamble only when A occurs? In this case, of course, you are guaranteed to win! What information can or cannot be used when setting up the decision series? Intuitively, as von Mises stated, your series of decisions must be *independent* of the collective. However, since the collective and the decision series are mathematical entities, the mathematical concept of independence needs to be defined first. But Axiom 2 is introduced to get independence in the first place; so again we face a circularity problem. Mathematicians in the Vienna circle developed a rigorous method to avoid that (Gillies, 2000), but the method does not seem to have any relevance in other areas of probability, so we will not go through it here.

We can show that Axiom 1 implies the finitely additive probability axioms, but it is not sufficient to guarantee countable additivity. Axiom 2 is not needed in FA probability, since randomness is not part of the standard probability axioms. Von Mises (1964) later added another axiom to cover CA, but this could not elevate von Mises's operationalist theory into a viable and attractive mathematical theory of probability.

So, it seems that we have arrived at a linguistic quandary: most philosophers attribute the frequency theory of probability to von Mises, but it is doubtful if there is any frequentist statistician that actually subscribes to the theory as developed by von Mises. It seems the frequentist statisticians simply follow Venn's early formulation of probability as a long-term frequency, not any full-blown axiomatic version of the theory. But without an axiomatic *operational* foundation, frequency interpretation can only be informal.

The LLNs cannot be used as the formal basis to define or interpret probability, since the statements of the theorems themselves need probability, so it will be a circular definition. The SLLN version guarantees convergence with *probability* 1. In the WLLN version, which corresponds to Bernoulli's Golden Theorem, the sample proportion $\widehat{\theta}$ converges to the true probability θ in the sense that, for any fixed $\epsilon > 0$ (think of a small value, e.g., 0.01), and *probability* p (think of a large value, e.g., 0.95), we can find a sample size n large enough such that we are guaranteed

$$\Pr(|\widehat{\theta} - \theta| \leq \epsilon) \geq p. \tag{6.1}$$

That is, with *high probability*, sample proportion can be made close to the true probability by taking a large enough sample. But how do we interpret this new probability p? Okay, let's imagine second-layer repeated experiments, each of size n and each time we check if $|\widehat{\theta} - \theta| \leq \epsilon$. Again, the WLLN implies we have *another* probability statement guaranteeing that the second-layer experiment will produce an observed proportion close to p, and so on. We are in a vicious circle. An easy way to break the circle is to take an epistemic interpretation of (6.1)! But that's of course an anathema for the frequentist.

6.4 Propensity Theory

The propensity theory was first formally formulated by Popper (1957), but later developed in several directions by other philosophers. We shall only describe the original formulation, which is actually adopted by many frequentist statisticians as part of their philosophy without attributing it to Popper's propensity theory.

As an objectivist, Popper originally supported the frequency theory, but he became unhappy with it, particularly with its inability to provide an objective probability of single events. Popper considered the following thought experiment. Suppose one starts with a collective C based on throwing a regular die, except, on some throws, say number 100, a biased die is used – it has '4' on two faces. The question is: What is the probability Pr(4)? According to frequency theory, which relies solely on the collective, it must be equal to 1/6. But, intuitively, for the 100th throw, we should have $\Pr(4) = 1/3$. Popper's solution is that the collective should be based on repetitions of an experiment under the same conditions, i.e., the same die, the same method of throwing, etc. In other words, the collective cannot be just an arbitrary sequence.

But, rather than fixing the collective, Popper came up with a new interpretation distinct from frequency theory: Probability is a property of the generating mechanism, i.e., a propensity of the mechanism to 'produce sequences whose frequencies are *equal* [our emphasis] to the probability.' What matters is that the probability applies to an event that is repeatable under the same conditions. There is no more need for an explicit collective as in von Mises's theory.

Technically, there is, of course, an issue here, since empirical frequency and probability are distinct entities; we have to deal with convergence or the central limit theorem. But, remember that Popper was not a mathematician. Interestingly and somewhat confusingly, in 1933, Kolmogorov was already suggesting a similar propensity interpretation for his probability, while at the same time stating that he was using von Mises's work to judge the applicability of the theory of probability to real-world events. Frequentist statisticians interpret the confidence level, for instance, as a property of the procedure that generates the confidence interval. So, in this sense, they are using the propensity interpretation of probability.

Objective probability of single events

In statistical applications, the question of having an objective probability for single events is of fundamental and practical importance. If the answer is positive, then, for instance, the confidence level – as an objective probability – applies to the observed confidence interval. Frequentists are unanimous in their rejection of this notion. Throughout his scientific career, Fisher argued that his fiducial probability applied to the observed confidence interval, despite the understandable rejection by the frequentists, especially by Neyman.

If von Mises's collective is not necessary in propensity theory and only a set of repeatable conditions is, does that mean that the propensity probability can apply to single events? That is, there is no need to actually replicate, as long as the conditions can be thought of as replicable. This was Popper's original motivation. But the fundamental problem is that the propensity is attached to the set of conditions rather than to the event itself, and single events are typically too heterogeneous to follow the same set of conditions.

Let's consider a simple example: what is the probability that John survives, say, beyond the next 10 years? Okay, to be objective, what are his characteristics? He is a 40-year-old white male, married with 2 children. Suppose among males with such characteristics, the probability of living beyond age 50, given that they are alive at 40, is 99%. Does this statistical probability then apply to him? But suppose it is also known that, as a banker, he works 12-hour days, including weekends, smokes two packs of cigarettes daily, and his own father died of heart disease at age 43. Now the previous probability no longer feels appropriate to him.

This is known as the reference class problem (Reichenbach, 1949). No matter how carefully we define an individual case, there are always further characterizations we could make. Even Aristotle – in *Rhetoric*, Book I, Chapter 2: 1356b – already recognized this issue:

> None of the arts theorize about individual cases. Medicine, for instance, does not theorize about what will help to cure Socrates or Callias, but only about what will help to cure any or all of a given class of patients. This alone is business: individual cases are so infinitely various that no systematic knowledge of them is possible.

Almost two millennia later, Venn (1876) in his classic book wrote:

> It is obvious that every individual thing or event has an indefinite number of properties or attributes observable in it, and might therefore be considered as belonging to an indefinite number of different classes of things.

The Principle of the Narrowest Reference Class had been suggested to overcome the problem: We should define an individual case as finely as possible. This sounds sensible, but we can easily run into insurmountable practical problems. There are always more individual characteristics than there are available statistical probabilities. In such a situation, an individual can belong to multiple classes with distinct statistical probabilities, but no information is available about the intersection class. For instance, among 40-year-old male smokers, the survival probability beyond 50 is 0.98; but among 40-year-old males whose fathers died by age 50, the probability is 0.90. But no information is available for the group of 40-year-old male smokers whose fathers died by age 50. Which probability is relevant? In practice, there is always a limit in data availability.

A statistician's practical consideration is typically based on balancing the trade-off between bias and variability: using a narrower class reduces bias but increases variability, and vice versa, so the narrowest reference class may not be the best idea.

In statistical inference, the reference class problem appears in an area called conditional inference (Reid, 1988). Conditional inference is the frequentist statistician's method to make inference more relevant to the case at hand. This is in turn connected to Fisher's concept of recognizable or relevant subsets. As we discussed previously, in Fisher's view (1958), a probability $\Pr(A)$ of an event A is a meaningful measure of logical uncertainty – hence epistemic – only if there is no relevant subset for which we can specify a distinct probability statement conditional on it. Specifically, suppose R is a subset of the sample space. Then R is a relevant subset if

$$\Pr(A|R) \neq \Pr(A).$$

We might argue that if there is no relevant subset, then the probability applies to the single event. We shall return to this topic in Chapter 13.

From the beginning, Popper argued for the propensity probability as an objective probability for single events, where the single event represents an instantiation of repeatable conditions. His wish for a single-event probability persisted over time, but his view about the need for repeatable conditions evolved. Thus, this latter-day propensity depends on 'the whole physical situation' (Popper, 1990) or the complete state of the world at the time the event is produced. This propensity no longer allows replication at all, because there will always be some changes in the state of the world between experiments. Unfortunately, it turns propensity theory into a metaphysical concept.

Most philosophers appear to remain unconvinced about whether we can have an objective probability for single events. For instance, Howson and Urbach (1989) wrote that such a probability can only be subjective. Except for obvious applications such as in the game of chance, where the set of conditions is easily replicable, Gillies (2000) also considered this probability to be more reasonably treated as subjective.

6.5 Subjective Theory

In frequency theory, probability is declared to be meaningless when we talk about unique single events, such as Donald Trump's re-election in 2024. But people do bet on such specific events. This can only mean that they do have a probability that is not of long-term variety. Thus, subjective probability is a much wider concept than objective probability. The reasoning used in betting requires a probability that would apply to propositions and single events. Furthermore, since different people have different beliefs and temperaments, they may have different probabilities for the same proposition despite being provided with the same information. This is a hallmark of subjective probability, which makes it distinct from Keynes's conception of logical probability.

Coherency axiom

Subjective probability is not necessarily as arbitrary as one suspects. It is underpinned by a rationality requirement. Setting a rational bet or a coherent bet is reminiscent of the I-cut-you-choose method of splitting a cake fairly between two greedy individuals. Even children will immediately see its fairness: John cuts the cake and Patrick chooses first. It is in John's interest to cut it as equally as possible. Cutting the cake into unequal pieces would not be coherent because he would knowingly guarantee that he will lose, as Patrick will surely choose the bigger piece. How do we extend this trick to the probability of an arbitrary event, such as the re-election of President Trump?

Say, John has a probability of 0.25 for such an event. It is not important that he knows it himself, but only that he is prepared to act – to bet – on it. It is the action, which speaks louder than any words, that reveals the subjective probability. We set up the following betting game. John (as a player) has the chance to win \$100 from Patrick (as a bookie) if the event occurs, but he has to pay Patrick a price to play the game. How much would he pay? Paying more than \$25 would violate his sense of the probability of the event, so he will not do that. But he might be happy to pay, say, \$10. Of course, this will not reveal his probability.

Here is the trick: Patrick has the right to reverse their roles as a player and bookie, depending on the amount John decides to pay. If John decides to pay \$10, then (he worries) he runs the risk that Patrick will reverse their roles and instead make him pay \$100 in return for \$10, if the event occurs. So, to neutralize this risk, John should pay as much as possible, but as he would not be willing to pay more than \$25, he will arrive at the fair price of \$25. At this price, John should feel indifferent to the role reversal. Choosing any other number would run the risk of a loss. So, the probability might also be called the risk-neutral (not risk-free) probability.

Several assumptions are implicit in this idealized betting game. One is precision: in real life, John the player will not have a precise price in mind, so precision is a mathematical convenience. At his fair price, it is assumed that John would be indifferent between being a player or a bookie. In practice, he may have different preferences, i.e., the chance of winning \$100 feels different from the chance of losing \$100, even if they are balanced by the same price or compensation.

A real-life Patrick the bookie will have his own probability of the event and will then set a price higher than that. For example, he may accept even \$10 if that is higher than his probability. So, the revelation of John's probability requires that Patrick not have his own price, while John must think introspectively and reflexively that Patrick is thinking like himself so that John's perceived risk of reversal occurs when he is willing to pay less than \$25. This condition is satisfied, for example, if both assume that they have access to the same background information.

Furthermore, it must be assumed that the amount of money involved is small relative to their wealth, so they can still think linearly. In large amounts, the utility of money becomes non-linear, so the reasoning must account for risk aversion, as people feel differently about losing \$10 vs \$10,000. Finally, the game might feel contrived with its potential role reversal and various assumptions. Role reversal means that you can be either the buyer or the seller of a bet. This is actually what happens in online betting exchanges, where you can choose to buy or sell depending on the difference between your subjective probability and the current 'market price' of an event. More generally, it corresponds to buying and 'shorting' (roughly means selling) an asset in the financial market.

Under those assumptions, the bets can be extended to an arbitrary but finite number of complementary events that partition the sure event, where John specifies the price for each event according to his subjective probabilities. The collection of bets constitutes 'a betting strategy.' The axiom of subjective probability is the following:

> **Coherency axiom:** No rational person will make a betting strategy that is guaranteed to lose money.

The counter-bet by an external agent that is guaranteed to win is called the Dutch Book, known more prosaically as 'free lunch,' or by its fancy name 'arbitrage' in the financial market. Any betting strategy protected from the Dutch Book is said to be *coherent*.

Imagine yourself running a bookie shop. For a specific event, say a boxing match between the heavyweight boxers Anthony Joshua and Tyson Fury, there are many potential bets you can sell. For instance, Joshua wins by any means, by KO, or by decision; Fury by any means, by KO, or by decision; a draw; or Joshua by KO in round 1, round 2, etc. There can easily be more than 25 different bets. All these bets must be priced so that no bettor can be guaranteed 100% to make money for free regardless of what happens in the match. In other words, you must have a coherent betting strategy.

Running a bookie shop in an open market is even more complex, since you will have competitors, but you also want to make sure that no bettor can take advantage of price differences between you and your competitors. This is made more difficult by the availability of instant information on the Internet. In reality, bookies protect themselves partly by setting a spread between buying and selling, which will translate into a cost for bettors who try to buy and sell bets on the same event. Moreover, bookies are also constantly on the lookout for anyone who wins too many bets; legally, they have the right to ban such a bettor. Since bookies presumably do not share information, how can they identify arbitrageurs? Each bookie must rely on their own information to analyse each player's long-term winning pattern, so there will be ways for clever arbitrageurs to avoid getting detected and permanently banned.

Ramsey-de Finetti theorem

Even though Keynes's idea of probability as a degree of rational belief did not have any direct impact in mathematics or statistics, it led Frank Ramsey in 1926, at the age of 22, to propose his alternative theory that views probability as a subjective degree of belief. It's an idea traceable to the Greek philosopher Protagoras (490BC-420BC?), that man is the measure of all things, typically interpreted to mean that there is no objective truth. Ramsey tried to develop a purely psychological method to measure belief. He argued that beliefs are like time intervals, and that they are somewhere within our introspective feelings, but that we can quantify beliefs by measuring the extent to which we are willing to act on them. During the 1920s, Bruno de Finetti also conceived the same idea independently of Ramsey, though his book was published only in 1937.

Thus, in their operationalist construction, Ramsey and de Finetti defined subjective probability as a betting price and proved their fundamental result:

> **Theorem:** A betting strategy is coherent if and only if it follows the axioms of (finitely additive) probability.

To illustrate, suppose that you set your betting prices as follows: (i) bet that the proposition G is true, risking α to win $1-\alpha$, and (ii) bet that G is not true, risking β to win $1-\beta$. Then I make two bets against you, on G and not-G simultaneously, either as a player or as a bookie depending on $(\alpha+\beta)$ below. My expected winning from the two bets is

$$\begin{aligned}&(1-\alpha)\Pr(G)-\alpha(1-\Pr(G))+(1-\beta)\Pr(\text{not } G)-\beta(1-\Pr(\text{not } G))\\&\quad=(1-\alpha)\Pr(G)-\alpha(\Pr(\text{not } G))+(1-\beta)\Pr(\text{not } G)-\beta(\Pr(G))\\&\quad=(1-\alpha-\beta)(\Pr(G)+\Pr(\text{not } G))\\&\quad=1-\alpha-\beta.\end{aligned}$$

If $(\alpha+\beta)<1$, I can always win by betting as a player, and if $(\alpha+\beta)>1$, I can always win by betting as a bookie. Therefore, to avoid a sure loss, your betting prices α and β must satisfy the additive probability axiom such that $(\alpha+\beta)=1$ and $\Pr(G)=\alpha\geq 0$ and $\Pr(\text{not } G)=\beta=1-\alpha\geq 0$. The resulting probability is subjective, because different people with the same information may have different betting prices.

We also get more from the coherency axiom. Since probability is operationally defined as a betting price, one can also define the probability of a joint event $A\cap B$ and the conditional probability in a similar way: $\Pr(A\cap B)$ is the betting price for the event $A\cap B$, and $\Pr(A|B)$ is the betting price for the event A, given that B is known to have happened. Then the coherency axiom implies that

$$\Pr(A\cap B)=\Pr(B)\Pr(A|B),$$

or

$$\Pr(A|B) = \frac{\Pr(A \cap B)}{\Pr(B)}.$$

This means that the conditional probability formula is implied by the coherency axiom, not by a definition as in the Kolmogorov system.

Savage's axioms

The implicit assumption of fixed linear utility renders the Ramsey-de Finetti construction insufficient as the basis of Bayesian statistics or Bayesian decision theory, where a general loss function – equivalent to minus utility – would be needed. The operationalist definition of probability as a betting price also makes it impossible to deal with the question of prior selection. With its rational preference-based non-operationalist abstract probability and general utility, it is Savage's work that established Bayesian statistics as a fully axiomatic deductive system.

With regular use of Laplace's inverse probability, 19th-century statistics was largely Bayesian in character, though not in name. But it wasn't until the 1950s that the Bayesian school of statistics emerged as a fully coherent statistical philosophy and methodology, perhaps conveniently coinciding with the publication of Savage's seminal book *The Foundation of Statistics* in 1954. Even the adjective 'Bayesian' got its current meaning after the appearance of many prominent Bayesians that revolved around Savage. Fisher (1930), following many 19th-century mathematicians, including George Boole and John Venn, criticized the Bayesian method for its arbitrary prior probability. Referring to this method, Fisher actually used the term 'Bayesian' in a pejorative sense.

There are two historical strands leading to Savage: (i) Ramsey-de Finetti (RDF) subjective probability as we describe above. (ii) von Neumann and Morgenstern's (VNM, 1947) subjective utility theorem in game theory that a person whose preferences follow certain rationality axioms behaves as if he is maximizing a utility function. In effect, the RDF theorem assumes objective and linear utility (good ol' money) and a coherent betting strategy to arrive at subjective probability. On the other hand, the VNM theorem assumes objective probability and axioms of rational preferences to arrive at subjective utility. Savage came up with the axioms of rational preferences that unify subjective probability and subjective utility within the decision theory framework. This axiomatic approach justifies the use of subjective probability, including subjective priors, in Bayesian statistics. Because of their fundamental importance, we shall cover Savage's axioms in full detail in Chapter 7.

It is interesting to see that in the philosophical literature, the subjective probability is dominated by Ramsey-de Finetti's idea. Philosophers appear to love the self-evident coherency axiom based on the Dutch Book argument. Whereas in the economics literature, the subjective probability is dominated by the much more mathematical – hence perhaps less self-evident – Savage's axioms or their variants. When a statistician claims to believe in subjective probability, presumably he believes in Savage's axioms, which are more complete than Ramsey-de Finetti's axiom.

Drawbacks of probability as degree of belief

There is no controversy over the use of probability for observable random events. However, when we talk about the use of probability for unobservable states of propositions or non-repeatable single events, it's a different story. In coin tossing, we can see the outcomes and probability is the long-term frequency of repeatable events that can be checked by observations. But the occurrence or non-occurrence of a single event does not verify its probability.

The propositions in epistemic probability can be single events in daily life. Thus, we may talk about the probability of rain tomorrow being 90% or the probability of economic recession next year being 75%, etc. There is a verifiability, deniability or falsifiability issue in such statements that is not in other objective measurements such as weight or height. When you say your weight is 80 kg, I can easily verify your statement by putting you on a weighing machine. But if you say that the probability of rain tomorrow is 90%, and the day turns out to be sunny, you can always say: Well, I'm not wrong, there is still a 10% probability of a sunny day.

When probability is applied to a space of propositions or hypotheses, naturally we have to enumerate all hypotheses such that 'probability 1' means that we are sure to cover all possibilities. To demonstrate the difficulty in achieving that, Senn (2003) described a jealous husband's dilemma. He fears that his wife might be unfaithful and hires a private detective to investigate her. After following her for weeks, investigating her phone calls and observing her every movement, he reports that there is no evidence that she is having an affair. 'The husband is much relieved.

He goes to bed with a light heart for the first time in months. At three in the morning, he wakes up with a frightening thought... suppose his wife is having an affair with the detective?' This thought was not in his mind a priori. If our actions change the world, then hypotheses could also be altered unexpectedly. Hiring a detective to investigate his wife could lead to the detective starting an unexpected affair with her. In such a world, we cannot enumerate all hypotheses a priori to cover all possibilities.

Probability does not exist?

De Finetti was not interested in finding the true state $I(G)$ of any proposition. According to him, we can simply issue the betting price $\Pr(G)$ as a personal epistemic degree of belief in a coherent way and use the degree of belief for actions. An extreme position one can take in regard to knowledge is that truth is not out there in the world, that truth cannot exist independently of the human mind. The world is out there, but descriptions of the world are not. Only descriptions of the world can be true or false. The world on its own, separate from humans' describing activities, cannot exist. For de Finetti, who shared this view of knowledge, $\Pr(G)$ was a personal belief, independent of the true state $I(G)$.

De Finetti (1974) was always emphatic in his rejection of objective probability. He declared 'paradoxically, and a little provocatively, but nonetheless genuinely,' that probability does not exist, in capital letters no less. The seemingly objective long-run frequency is for him an illusion because of agreements between the observers. This leads to our next section, where agreement between individuals is seen not as an illusion, but rather as an imagined reality created naturally by powerful human cognitive competence. It seems ironic to view probability as a betting price, yet not to see the existence of betting markets, where prices are agreed widely by different individuals without any need for long-term frequencies.

6.6 Consensus theory

According to subjective theory, two people with the same information don't necessarily have to have the same (subjective) probability. The reason is that they can choose to act differently. This might sound appealing to researchers in social studies, but less so to natural scientists. While the events studied in the social sciences are associated with thinking agents/participants, natural phenomena are not. The question is whether we can reach consensus among a group of rational independent people. Can such a group of people agree on a common betting price? This is what determines the action of the group.

Any probability assignment for coherent betting is imagined in the sense that it only describes a state of mind, and not something that can be measured in the physical world. However, this imagined status does not necessarily imply subjectivity. Suppose that a group of closely related individuals has to bet on an event. If the group members set different betting prices for the same event, then an outsider can take advantage by buying from the low-price member and selling to the high-price one. In the financial market this is known as arbitrage. The members' behaviour is irrational if the group has a common economy. This suggests an extension of the Dutch Book idea to the market setting, which we discuss below. For now, we just emphasize that there is pressure on the group to come up with a consensus price. This is the consensus probability; as we have discussed on page 68, a consensus price has an objective existence as an imagined reality. There are some conditions required for the market to achieve a consensus price (see the market equilibrium theory below), but the ubiquitous existence of markets everywhere around the world is evidence that the conditions are easy to achieve, not perfectly but adequately, in order to arrive at agreed prices.

There are different levels of non-subjectivity. A fully objective probability would refer to values that are completely independent of the human mind. We typically consider long-term frequencies as those objective probabilities. Yet, for the physical devices in the game of chance, including the coins, cards, dice, etc., even skeptical gamblers can agree on the commonly assumed probabilities at each game, not requiring long-term repetitions of the game. To make the game happen, it is enough that all gamblers agree on the probabilities. Thus, consensus probabilities are epistemic yet non-subjective, describing the group's state of mind that is independent of individual subjective preferences. Gambling on a horse race, on the other hand, does not rely on a physical device; bookies determine prices based on the past performances of the horses, which are known to all gamblers, so the prices are not arbitrarily subjective, but not fully consensual as we would expect for physical devices.

Objective Bayesianism

Subjective probability is operationalized when people make individual bets. Where do we see any working version of consensus probability? Let's look into the Bayesian world. Bayesian statistical inference, which we will cover in detail in Chapter 8, is marked by the use of prior probabilities. Subjective Bayesians interpret probabilities in terms of the odds at which someone is willing to bet, whereas objective Bayesians interpret them as rational degrees of belief given the available evidence. For a stereotypical subjective Bayesian, all probabilities are just opinions and what's important is coherence. Two subjective Bayesians with the same information don't necessarily end up with the same conclusion. In their thinking, each person is allowed to adopt one's own prior.

On the other hand, we suppose objective Bayesians expect all rational people with the same information to have the same prior and yield the same posterior. Most objective Bayesians use the prior to represent ignorance due to a lack of information, about which they believe some agreement should be reached. So we can recognize that their aim is to arrive at a consensus probability. It also shares the same goal as Keynes's logical probability.

Market-based Dutch Book argument

Non-Bayesians generally reject subjective probability, but would they accept a consensus probability? To avoid complete subjectivity, we need to extend the Dutch Book argument beyond two people. This idea will be heavily used in Chapter 13 on epistemic confidence. Let's assume that there is a betting market involving a crowd of independent and intelligent players. In this market, bets are like a commodity with supply and demand between players. Assuming a perfect market condition – e.g., full competition, perfect information and no transaction cost – by the Arrow-Debreu theorem (Arrow and Debreu, 1954), there is an equilibrium price at which there is balance between supply and demand. Intuitively, if you are a seller and you set your price too low, many would want to buy from you, thus creating demand and increasing the price. However, if you set your price too high, nobody wants to buy, thus increasing the supply and pressing the price down. Theoretically, there is of course no need for a step-by-step process to converge to an agreed price; in a perfect market, the intelligent buyers and sellers are supposed to find the equilibrium price instantly.

For the betting market in particular, the fundamental theorem of asset pricing (Ross, 1976) states that, assuming an objective probability model, there is no arbitrage if the price is determined by the objective probability. (If there is no objective probability model, the market – as evidenced by actual betting markets – can still have an agreed price at any given time, though not a theoretically determined price.)

To illustrate the role of the betting market in the Dutch Book argument, suppose that you and I are betting on the 2024 US presidential election. Suppose that the betting market is currently offering the price 0.25 for Donald Trump to win (this means you pay \$0.25 to get \$1 back if Trump wins, including your original \$0.25). On the other hand, for whatever reason, you believe that Trump will lose, and hence set the probability of him winning at 0.1. Then I would construct a Dutch Book, by buying from you at 0.1 and immediately selling it on the market at 0.25, thus making a risk-free profit.

'Buying from you' means treating you like a bookie: paying you \$0.1 and getting \$1 back if Trump wins. While 'selling in the market' for me means betting against the event, so I behave like a bookie: people pay me \$0.25 so that they can get \$1 back if Trump wins, but I keep the \$0.25 if Trump loses. So, overall, I would make \$0.15 risk-free, i.e., regardless of whether Trump wins or loses. Note that this is not just a thought experiment: You can do all this buying and selling of bets in the online betting market, though in the real market there are of course transaction costs.

It is worth highlighting the difference between the market-based Dutch Book argument and the classical version used to establish subjective Bayesian probability. In the classical setup, because it does not presume the betting market, multiple bets are made only between you and me. To avoid the Dutch Book, you have to make your bets internally consistent by following probability laws. However, *even if* your bets are internally consistent, as illustrated above, as long as your prices do not match the market prices, I can make a risk-free profit by playing between you and the market.

So, ignorance of the market creates a potentially different kind of sure loss, which economists call an opportunity cost. You either lose the opportunity to make more money by selling directly to the market (if your price is lower than the market price), or you lose the opportunity to get lower prices by not buying directly from the market (if your price is too high). Therefore, the consensus probability can also be justified by a more general coherency axiom that no one will set a betting strategy that is guaranteed to make a loss, including the loss of opportunity.

Objective probability of single events

Finally, let's come back to the issue of objective probability of single events. In the market, there is an objective price for each specific item, a price accepted by the crowd of market participants. This means that the market-based Dutch Book allows construction of objective probability as a betting price for single events. How do we avoid the sticky reference class problem? The key is the perfect market hypothesis, which assumes, among other things, perfect information for everyone in the crowd, so in principle, they all use the same reference class. In stock markets, the equal access to information by all market participants is enforced by law; the use of insider information is considered criminal.

How do we operationalize the consensus probability for single events in the non-Bayesian world, considering that the conceptual use of 'probability' for single events will always be controversial? This is getting ahead of ourselves, but it is indeed possible using the concept of epistemic confidence, which we discuss in Chapter 13.

7

Rereading Savage

If we accept Savage's axioms (Section 6.5), then all uncertainties must be treated as probability. Savage espoused subjective probability, allowing, for example, the probability of Donald Trump's re-election in 2024. But Savage's probability also covers the objective version, such as the probability of heads in a fair coin toss. In other words, there is no distinction between objective and subjective probability. Savage's system has great theoretical implications; for example, prior probabilities can be elicited from subjective preferences and then updated by objective evidence, a learning step that forms the basis of Bayesian computations.

Non-Bayesians have generally refused to accept the subjective aspect of probability or allow priors in formal statistical modelling. As demanded, for example, by the late Dennis Lindley, since Bayesian probability is axiomatic, it is the non-Bayesians' duty to point out which axioms are not acceptable to them. This is not a simple request, since the Bayesian axioms are not commonly covered in standard statistical training, even in Bayesian statistics courses. Therefore, our aim is to provide a readable exposition of Bayesian axioms from a close reading of Savage's classic book, *The Foundation of Statistics.* It was first published in 1954, but we used the 1972 version. His work established Bayesian statistics axiomatically, bringing mathematical legitimacy to prior probability and subjective probability in general.

Savage's axiomatic development is a culmination of the subjective theory of probability at the heart of Bayesian statistics, so it seems appropriate to put our focus on him. Rather than 'subjective', Savage liked the term 'personalistic,' but the latter is not in common use, so we shall use 'subjective' throughout.

7.1 The Axioms

Savage conveniently listed his seven axioms as the Postulates of a Personalistic Theory of Decision at the front of his book. The 'formal subject matter' includes

(i) S the set of 'states of the world' $s_1, s_1', \ldots,$

(ii) F the set of 'consequences' $f_1, f_1', \ldots,$

(iii) the set of 'acts' $\mathbf{f}_1, \mathbf{f}_2, \ldots.$

Formally, an act $\mathbf{f}$ is a function from S to F. The function $f(s)$ is the consequence at state s, but $\mathbf{f}$ refers the whole function $f(\cdot)$ across s. To facilitate understanding – and to tie up more easily with Ramsey and de Finetti (RDF), and von Neumann and Morgenstern (1947, VNM) – think of an 'act' as a bet on the state of the world, and the 'consequence' as the payoff. (In general, the payoff can be expressed on a utility scale, but there is no need to do that yet.)

The 'state of the world' is a description sufficient to compute the consequences of each act. It is useful to envision the usual random experiment with S as the sample space, but, unlike in von Neumann and Morgenstern's lottery, don't assume that the random generating mechanism is known. For example, imagine betting on a horse race: The bet and the payoff are obvious, and the sample space is the list of horses. The process of producing a winner is a complex random process; Savage's theory expresses a person's probability model in terms of his betting behaviour.

An abstract term 'state space' – rather than 'sample space' – is used, as the theory makes no distinction between the uncertainty due to ignorance, i.e., involving unknown fixed parameters, and the uncertainty due to true randomness. In the case of fixed parameters, pedagogically, they are usually presented as if their true values were sampled from a subjective probability distribution, but formally, no sampling is needed.

Allowing both uncertainties in a single framework creates linguistic challenges. A state is said to 'obtain' if it is the 'true state of the world,' meaning the true parameter in the parameter space S, or 'realized' if it is a randomly generated state in the sample space S. For example, the winning horse is the true state among the participating horses in a race. The subsets of S are called events. An event is also said to 'obtain' if it contains the true state; in standard probability terms, we say that the event 'occurs.' We shall use the latter term, because the expression 'an event obtains' sounds awkward no matter how many times we hear it. When the state is a parameter, imagine the true state was generated randomly, so the word 'occur' also makes sense.

TABLE 7.1

A race of three horses A, B and C, and the corresponding payoffs associated with the bets $\mathbf{f}_1$, $\mathbf{f}_2$ *and* $\mathbf{f}_3$.

Bet	Horse A	Horse B	Horse C
$\mathbf{f}_1$	\$100	\$0	\$0
$\mathbf{f}_2$	\$0	\$100	\$0
$\mathbf{f}_3$	\$0	\$25	\$25

Table 7.1 shows a simple example that captures the elements of the theory. Choosing the bet $\mathbf{f}_1$ means that you believe Horse A would win, or more realistically, horse A has a higher chance of winning than the other horses. Similarly, choosing $\mathbf{f}_2$ means you believe in Horse B; choosing $\mathbf{f}_3$ means you believe either B or C would win. As in Ramsey's and de Finetti's theory, a betting choice reveals a subjective probability. In one key difference with the standard probability model: A horse race may already be finished. As will be clear from the axioms, the whole theory applies unchanged to the bets made on the result of a specific race, even a race that has concluded as long as the result remains unknown to the bettor. In contrast, the frequentist probability does not apply to the result of a specific race. We can see here the attraction of subjective probability, as people do bet on specific events.

The final element of the theory is a preference ordering between the acts: '$\mathbf{f}_1 \leq \mathbf{f}_2$' means '$\mathbf{f}_1$ is not preferred to $\mathbf{f}_2$' or, more simply, '$\mathbf{f}_2$ is preferred to $\mathbf{f}_1$; this is equivalent to $\mathbf{f}_1 < \mathbf{f}_2$ or $\mathbf{f}_1 = \mathbf{f}_2$, meaning either $\mathbf{f}_2$ is strictly preferred over $\mathbf{f}_1$ or they are equivalent. Now we are ready for Savage's seven postulates.

Axiom 1

P1 (Weak ordering) The preference order is complete and transitive, meaning that (i) every pair $\mathbf{f}_1$ and $\mathbf{f}_2$ can be compared and ordered, and (ii) for every $\mathbf{f}_1$, $\mathbf{f}_2$ and $\mathbf{f}_3$, if $\mathbf{f}_1 \leq \mathbf{f}_2$ and $\mathbf{f}_2 \leq \mathbf{f}_3$, then $\mathbf{f}_1 \leq \mathbf{f}_3$.

This postulate is similar to VNM's Axioms 1 and 2 for subjective utility. Transitivity is a crucial element of any axiomatic system of rational behaviour. A typical concern is on completeness: we can imagine situations where we put little thought into, hence have no explicit preferences over, some acts. In statistics, this is not likely to be a concern, and Savage advised against considering a system that allows partial ordering. As a simple consequence of the axiom, if F is finite, then there must be an optimal act, i.e., no other act is strictly preferred over it.

From the beginning, Savage had already pointed out two aspects of an axiomatic system: empirical and normative. The former is a descriptive or predictive theory of people's behaviour in decision making: The axioms will be considered good or bad according to the closeness of the predictions to the actual behaviour of people. The normative aspect attempts to guide people's behaviour, particularly behaviour that is not consistent with the theory. He brought analogy to logic, where once we agree to certain propositions, then we should follow their logical consequences. Axiomatic systems such as Savage's or VNM's derive their supposed normative values from this iron law of logic. Savage's correction of his own reaction to Allais's paradox (Section 14.1) shows that he considered his system normative. So in this normative sense, Savage's axioms are axioms of rational behaviour.

Axiom 2

> P2 (Sure-thing principle) For every act $\mathbf{f}, \mathbf{g}$ and event E, $\mathbf{f} \leq \mathbf{g}$ given E or $\mathbf{g} \leq \mathbf{f}$ given E.

This axiom at the front-end of the book looks deceptively simple, raising the question of why it is called 'the sure-thing principle' and why it is controversial. The idea of conditional preference as used in the axiom requires care, because different pairs of acts might agree when E occurs, but do not agree when E does not occur (denoted by E^C). Conditional preference is defined as follows:

> $\mathbf{f} \leq \mathbf{g}$ given E, if and only if $\mathbf{f}' \leq \mathbf{g}'$ for every $\mathbf{f}'$ and $\mathbf{g}'$ that agree with $\mathbf{f}$ and $\mathbf{g}$, respectively, on E, and with each other on E^C and $\mathbf{g}' \leq \mathbf{f}'$ either for all such pairs or for none.

Now that does not look so simple. Within the text, the postulate gets an alternative longer version, not requiring a definition:

P2 (Sure-thing principle) If $\mathbf{f}$, $\mathbf{g}$, and $\mathbf{f}'$, $\mathbf{g}'$ are such that:

1. in E^C, $\mathbf{f}$ agrees with $\mathbf{g}$, and $\mathbf{f}'$ agrees with $\mathbf{g}'$,
2. in E, $\mathbf{f}$ agrees with $\mathbf{f}'$, and $\mathbf{g}$ agrees with $\mathbf{g}'$;
3. $\mathbf{f} \leq \mathbf{g}$;

then $\mathbf{f}' \leq \mathbf{g}'$.

Using a definition in one place and an axiom in another is an interesting stylistic choice; in the latter, the conditional preference is an undefined primitive concept characterized in the axiom. In case you find this longer version a bit challenging, Savage illustrated it with an example (Savage, 1972, page 21):

> A businessman contemplates buying a certain piece of property. He considers the outcome of the next presidential election relevant to the attractiveness of the purchase. So, to clarify the matter for himself, he asks whether he would buy if he knew the Republican candidate were going to win, and decides that he would do so. Similarly, he considers whether he would buy if he knew the Democratic candidate were going to win, and again finds that he would do so. Seeing that he would buy in either event, he decides that he should buy.

He then added famously that 'I know of no other extralogical principle governing decisions that finds such ready acceptance.' That's why he called the postulate the sure-thing principle. Yet, mapping the example to the postulate is not obvious, because the postulate involves four distinct acts, while the businessman contemplates only two. In fact, the example is a more appropriate illustration of the so-called dominance principle, which involves consistent ordering of two acts across the states of the world.

Let's consider instead the example in Table 7.2 involving four bets in a horse race, which represent the bets on A, B, {A,C} and {B,C} respectively. Define the event $E \equiv$ {A,B}. The pairs of bets $(\mathbf{f}, \mathbf{g})$ and $(\mathbf{f}', \mathbf{g}')$ agree on E, while on the complement event E^C the bets agree within each pair. In words, if E does not occur, the two bets within each pair are equivalent; while if E occurs, the two pairs are indistinguishable in their payoffs. Thus P2 specifies that if, given E, you decide $\mathbf{f} \leq \mathbf{g}$ then you must also decide $\mathbf{f}' \leq \mathbf{g}'$, and vice versa. This seems reasonable: In the within-pair comparisons of the bets, only horses A and B contribute preferential information, and C is irrelevant because it produces

TABLE 7.2

A race of three horses A, B and C, and the corresponding payoffs associated with four bets labelled according to the acts in postulate P2. The pair (f, g) is a choice between A vs B, while the pair (f′, g′) between not-B vs not-A. So the two pairs are distinct choices, but, because C has the same payoff within each pair, according to P2 the preference order within each pair must be the same.

Bet	Horse A	Horse B	Horse C
f	\$100	\$0	\$0
g	\$0	\$100	\$0
f′	\$100	\$0	\$100
g′	\$0	\$100	\$100

the same payoff within each pair. However, Allais (1953) and Ellsberg (1961) proposed paradoxes that question the empirical validity of this axiom; see Chapter 14.

Versions of the postulate P2 appear in other axiomatic systems. It is closely related to Rubin's postulate or Milnor's column linearity postulate (Luce and Raiffa, 1957), which states that adding a constant amount to a column of payoffs – associated with adding one state you don't care about – should not change the ordering between acts. If the probabilities of the events are known, as in a lottery, P2 is equivalent to VNM's fourth axiom, called the independence axiom, stipulating that the decision maker compares alternatives based on their distinct consequences, and ignores aspects that are the same. In social choice theory (e.g., Arrow, 1950), the axiom is related to the axiom of independence of irrelevant alternatives.

Axiom 3

> P3 (State independence) If $\mathbf{f} \equiv g$, $\mathbf{f}' \equiv g'$, and event E is not null, then $\mathbf{f} \leq \mathbf{f}'$ given E, if and only if $g \leq g'$.

First define the preference ordering of consequences g and g' in terms of constant acts $\mathbf{f} \equiv g$ and $\mathbf{f}' \equiv g'$; think of these acts as receiving gifts instead of betting, since you get paid no matter what happens. The axiom states that the ordering of consequences remains the same under any event, or that knowledge of the underlying event does not change the preferential ordering of consequences. P3 is followed by a significant theorem: Assuming axioms P1 and P2, axiom P3 is equivalent to the admissibility or dominance principle, a key principle in Abraham Wald's statistical decision theory. In brief, an act is not admissible if there is another act that dominates it in terms of its consequences. The principle states that we must reject non-admissible acts: Roughly, these are bets whose payoffs at each state are no better than those of another bet, and strictly worse for at least one state.

Leading to P3, Savage also defined a null event, which is an event under which all acts are equivalent; for example, when all bets have the same payoff, so the decision maker could ignore it from further preferential considerations. According to this definition, a null event is *not* an impossible event in the usual probabilistic sense, but one about which the decision maker does not care enough so that it makes no impact on his preferences. For example, for someone who does not care to write his will and has no preference for what happens to his wealth after he dies, all the life-threatening events are null events. It also means that for the states that matter, i.e., all the non-null states, there must be at least one consequence that is distinct from the others. This will appear in postulate P5.

Since, in the limit, an event can contain a single state, the axiom can be called ordinal state independence of payoffs. Thus, the preferential value of \$100 payoff is the same regardless of the underlying state that produces it. This seems appropriate for statistical applications, but there are real-life situations where this may not be the case. In general, the axiom leads to state-independent and chance-neutral utility. In health economics or insurance applications, the utility of money changes depending on a person's underlying state of health. Chance neutrality means, the value one feels about, say, \$1000 is the same whether it comes from the regular salary or from winning the lottery. In reality, the latter is often called 'hot money,' which is spent more easily than hard-earned money. Recently Stefánsson and Bradley (2015) discussed this aspect in rationality and utility theory.

Axiom 4

P4 (Consequence independence) If consequences f, f', g, g'; events A, B; and acts $\mathbf{f}_A, \mathbf{f}_B, \mathbf{g}_A, \mathbf{g}_B$ are such that:

1.	$f' < f,$	$g' < g$	
2a.	$f_A(s) = f,$	$g_A(s) = g$	for $s \in A$
	$f_A(s) = f',$	$g_A(s) = g'$	for $s \notin A$
2b.	$f_B(s) = f,$	$g_B(s) = g$	for $s \in B$
	$f_B(s) = f',$	$g_B(s) = g'$	for $s \notin B$
3.		$\mathbf{f}_A \leq \mathbf{f}_B;$	

then $\mathbf{g}_A \leq \mathbf{g}_B$.

This looks pretty complicated, but actually not. The act $\mathbf{f}_A$ can be interpreted as a bet on event A, as it offers a strictly better payoff if A occurs; similarly for the other three acts. So $\mathbf{f}_A \leq \mathbf{f}_B$, i.e., preferring B over A even though they have the same payoffs, must mean you believe B is more likely than A. The axiom states that the choice of events to bet on is independent of the size of the payoff. In the horse race example (Table 7.1), the choice of Horse B over A should remain the same if the payoff is changed to \$200 or any other value.

Axioms 3 and 4 are the key pillars that support a great weight: the separation between subjective probability and utility, or between belief and value, which are normally entangled in ordinary preferences. This means (i) utility of payoffs can be defined independently of the underlying state, including its probability, and (ii) the subjective probability of an event is meaningful independently of the value of the event. The former can also be interpreted as chance neutrality: When you win, say \$1M in a lottery, you should feel the same regardless of whether the winning probability is close to impossible 10^{-18} or just small 10^{-6}.

Axiom 5

P5 (Non-triviality) There is at least one pair of consequences f, f' such that $f' < f$.

This must be the case at least under one state of the world. In view of the discussion of null events in Axiom 3 above, this axiom is needed to avoid a trivial null state space. In other words, there is at least one event that you are willing to bet on.

7.2 Qualitative Probability and Axiom 6

Axioms 1–5 form a natural set, as they are sufficient to construct a qualitative probability. It corresponds to the logical probability concept that appeared in Keynes's proposal (Section 6.2), which he considered necessary when there is not sufficient knowledge to set up quantitative probability. In this framework, we can say, for example, that event A is less probable than B without specifying their numerical probabilities. It also corresponds to the qualitative probability in Polya's (1954) plausible reasoning.

Defining 0 as a null event, S as the state space, a relational operator $\preceq$ is a qualitative probability if, for all events B, C and D,

1. $\preceq$ is complete and transitive,
2. $B \preceq C$ if and only if $B \cup D \preceq C \cup D$, provided $B \cap D = C \cap D = 0$,
3. $0 \preceq B$, $0 \preceq C$.

It is, however, more convenient to express the relation in probability notation. The probability measure Pr and the qualitative probability $\preceq$ are said to agree if, for every B and C, $\Pr(B) \leq \Pr(C)$ is equivalent to $B \preceq C$. Unfortunately, Axioms 1–5 are not sufficient to guarantee the existence of a probability measure that agrees with $\preceq$. Savage proved a series of theorems leading to postulate P6′, which provides the guarantee.

> P6′ If $B \prec C$, there exists a partition $\cup_i E_i$ of S, such that $B \cup E_i \prec C$ for any i.

The key idea here is that S must be rich enough so that it can be divided into a fine partition, where any piece of the partition is too small to change the strict ordering of B and C. Savage viewed this postulate as a precursor to postulate P6 needed for a full-blown quantitative subjective probability.

> P6 (Small-event continuity) If $\mathbf{f} < \mathbf{g}$, and h is any consequence, then there exists a partition $\cup_i E_i$ of S such that, if $\mathbf{f}$ is so modified to take value h on E_i for any i, other values being undisturbed,

then the modified $\mathbf{f} < \mathbf{g}$. The same ordering is also preserved when $\mathbf{g}$ is modified instead.

The postulate requires S to be rich enough, at least infinitely countable, to allow a fine partition, and each act is a smooth function on S so that a modification of a small event does not disturb the preference of the act. Typically, this is interpreted to mean there is no consequence, either infinitely better or infinitely worse than the others, for which its inclusion in a modification of an act upsets its ordering. Another consequence is that the probability is *non-atomic* in the sense that for any ρ, $0 \leq \rho \leq 1$, and event E, there is $B \subseteq E$, such that $\Pr(B) = \rho\Pr(E)$.

Allais's paradox in Section 14.1 shows that a small-event discontinuity can occur, even with bounded consequences, if the small event turns an act into a sure bet. The continuity axiom is closely related to VNM's third axiom, which, for no obvious reasons, is called the Archimedean property. It is a continuity requirement in the set of acts – lotteries – in that no act is infinitely more or infinitely less preferable than any other act. Further technical conditions are needed to deal with finite state spaces as in the horse race example.

Axioms 1–6 are sufficient to construct quantitative/numerical subjective probability, which conveniently reduces the calculation of preferences between acts in terms of arithmetic comparisons of numbers. However, the acts must be of a special form, i.e., they are bets on events. So what we have is the numerical probability of any event. At this point, the theory is difficult to navigate, since there is no explicit theorem stating exactly what has been achieved by the time we reach Axiom 6. Comparisons of arbitrary acts with arbitrary consequences will have to wait until the introduction of utility and Axiom 7.

Finitely additive probability

Similar to RDF's coherent betting strategy, Savage's axioms 1–6 are necessary and sufficient to construct FA probability. This is in contrast to Kolmogorov's countably additive (CA) probability, which is the basis of most results in modern probability theory. There are, however, modifications of Savage's theory – for instance, Villegas's (1963) monotonicity condition – that lead to CA probability. We compared FA probability to CA probability in Section 5.2.

7.3 Conditional Probability

Conditional probability can be constructed starting with conditional preference as described in Axiom 2. If $\preceq$ is a qualitative probability relation, then for events B and C, and non-null event B, we can define the conditional relation $B \preceq C$ given D to mean $B \cap D \preceq C \cap D$. It can then be shown that the conditional relation is also a qualitative probability. If $\preceq$ is numerically represented by (or almost agrees with) the probability measure $\Pr(\cdot)$, then the conditional relation is represented by

$$\Pr(B|D) = \frac{\Pr(B \cap D)}{\Pr(D)}.$$

This is of course the key formula that allows updating of subjective probability, the crucial step of learning from experience that forms the basis of Bayesian inference. Savage then moved imperceptibly from the qualitative to a full quantitative conditional probability; presumably all follows from Axioms 1–6.

Then came (i) a familiar introduction to the Bayesian method with updating of prior probability $\Pr(B_i)$ of event B_i, using data $x \equiv (x_1, \ldots, x_n)$, to produce posterior probability

$$\Pr(B_i|x) \propto \Pr(x|B_i)\Pr(B_i)$$

and (ii) an exposition of exchangeability based on de Finetti's example on the inference of the success probability p from a sequence of Bernoulli trials. The classical objective view is that p is a fixed unknown parameter. Agreeing with de Finetti, 'objectivity' is seen simply as an agreement of subjective opinions. This is superficially in line with the idea that consensus has an objective imagined reality. However, for them, different people may start with different opinions – hence different priors – but their posterior probabilities will eventually agree, so there is no need for an objective view. In both expositions, there is an emphasis on asymptotic consistency as n tends to infinity, and there is no discussion of how to obtain the prior distribution.

7.4 Utility, Axiom 7 and The Theorem

Up to this point, acts that are bets on events can be compared using numerical probabilities. But we cannot compare acts with arbitrary consequences, which is, of course, necessary for general statistical applications. To do that, we need the concept of expected utility $\mathcal{U}$, such that

$$\mathbf{f} \leq \mathbf{g} \text{ if and only if } \mathcal{U}(\mathbf{f}) \leq \mathcal{U}(\mathbf{g}). \tag{7.1}$$

In this case we say that the preference ordering agrees with the utility. Given a probability measure $\Pr(\cdot)$ on S, the expected utility $\mathcal{U}(\mathbf{f})$ is

$$\mathcal{U}(\mathbf{f}) \equiv \int_S u(f(s))d\Pr(s), \tag{7.2}$$

where $u(\cdot)$ is the utility function applied to consequences $f(s)$. The utility concept allows a convenient arithmetic representation of acts and their preference ordering. If we consider the axioms normatively as axioms of rational decisions, then the utility function is a calculator of rationality, so it has a special role in any discussion of rationality and economic behaviour.

Savage's utility theory is largely influenced by VNM's utility. In the final foundational chapter (Chapter 5), Savage discussed the utility concept at great length, providing the historical background from the time of Daniel Bernoulli in the 18th century, and defending it from the then orthodox economic view that had dismissed any meaningful value of cardinal – as opposed to ordinal – utility. He was siding strongly with von Neumann and Morgenstern, and even going beyond them by interpreting the theory normatively as a guide to rational behaviour.

As we have seen above, three of Savage's six axioms are in fact closely related to VNM's four axioms to establish the existence of subjective utility as the basis of preferential ordering of lotteries. By first defining a gamble as an act with a finite number of consequences, Savage first proved that Axioms 1–6 are sufficient to guarantee the existence of a utility function that agrees with the preferences between gambles. So Axiom 7 is introduced to allow for a theoretically infinite number of consequences:

> P7 (Strong dominance) If $\mathbf{f} \leq g(s)$ for every s in B, then $\mathbf{f} \leq \mathbf{g}$ given B.

This means that if every consequence of **g** is preferable to **f** as a whole, then **g** must be preferable to **f**. Savage considered this as a stronger version of the sure-thing principle. P7 is clearly a stronger version of the dominance principle, which only compares the consequences $f(s) \leq g(s)$ in each state s. As we mentioned above, Savage proved that, assuming P1 and P2, postulate P3 is equivalent to the dominance principle. This means that P7 makes P3 redundant in the whole set of axioms (cf. Hartmann, 2020).

The Theorem

The utility concept completes Savage's axiomatic development, and the overall theorem is stated as follows: The preference ordering of acts satisfies the postulates P1–P7 if and only if there exists a unique finitely additive non-atomic probability measure P on S and a real-value bounded utility function $u(\cdot)$ on F, such that the preference ordering agrees with expected utility in the sense of (7.1).

This means someone who acts rationally, in the sense that his preferences satisfy the axioms, behaves *as if* he has a probability measure and a utility function, and he maximizes the expected utility. In principle, one does not even have to be aware of his subjective probability or utility function. However, in practice, it is much easier to first assume explicitly the probability and the utility function and then work out the implied preferences. For example, in Bayesian statistics, one typically starts with a prior distribution and a minus squared-error loss as utility. According to the theorem, maximizing the assumed expected utility corresponds to choosing an optimal act. By staying within the convenient utility framework, or, in other words, 'by following the expected utility theory,' your preferences are guaranteed 'rational.' Any decision that violates the expected utility can be shown to violate at least one of the axioms.

Savage's theorem can be seen as an amalgamation of Ramsey-de Finetti's subjective probability and VNM's subjective utility. Strictly speaking, when the probability is objective – as in a lottery – the utility theory is covered by VNM's framework, otherwise it is in Savage's.

7.5 Discussion

Savage dedicated a whole chapter to addressing criticisms of the (personalistic) subjective probability and to giving his own criticisms of other views of probability. First, it is perhaps useful to go through a number of statements that capture his views (pages 46, 56, 57). We can recognize here his strong influence on modern Bayesianism as an all-embracing theory, expressed, for example, by Lindley (2000):

> ...any mathematical problem concerning personal probability is necessarily a problem concerning probability measures – the study of which is currently called by mathematicians mathematical probability – and conversely.
> ...the concept of personal probability introduced and illustrated in the preceding chapter is... the only probability concept essential to science and other activities that call upon probability.
> ...the role of mathematical theory of probability is to enable the person using it to detect inconsistencies in his own real or envisaged behavior.

Ambiguous probability

Ellsberg's paradox (Section 14.2) highlights an example where people react differently to different types of uncertainty, roughly as 'sure' and 'unsure' or ambiguous probability. Savage dismissed the notion of second-order probability to deal with such different levels of uncertainty as in 'the probability that B is more probable than C is greater than the probability that F is more probable than G.' The concept of ambiguous belief or 'imprecise probability' was later taken up, for example, by Gilboa and Schmeidler (1993) and Binmore (2009).

Savage also did not support a model in which the probability of B is a random variable with respect to the second probability. He was concerned with 'endless hierarchy', which would be difficult to interpret. Hierarchical models of this type have in fact become popular in recent years. It is now common to think of Bayesian models as hierarchical, but Savage did not consider subjective probability as part of a hierarchical model.

Inexact magnitude

As with RDF's coherency argument, the postulates imply that one can determine with high accuracy, for instance, his subjective probability that Donald Trump will be re-elected. In reality, we can only expect some rough magnitudes. Savage responded that the subjective theory, as a normative theory, is 'a code of consistency for the person applying it, not a system of predictions.' De Finetti's (1931, page 204) response to the same issue is also relevant here: 'to apply mathematics, you must act as though the measured magnitudes have precise values. This fiction is very fruitful... To go, with the help of mathematics, from approximate premises to approximate conclusions, I must go by way of an exact algorithm ... .'

Other views of probability

Savage considered the subjective probability as lying not in between, but beside, the logical and objective views, because the latter two are meant to be free of individual preferences. For him, the strength of his axiomatic system is that it explicitly deals with the problem of making a decision under uncertainty. He saw a weakness in other subjective views, such as Koopman's (1940), that did not explicitly deal with the problem of individual decisions. De Finetti's coherent betting approach was criticized because it needs to assume a linear utility function (or the bet is small).

For the objective view – i.e., frequentist – probability is primary, decision secondary. Furthermore, probability is only given to very special events, e.g., repeatable ones such as coin tosses, but not to unique ones such as the unification of Korea. For Savage, the objective view is 'charged with circularity:' It relies on the existence of processes that can be represented by infinitely repeatable events, but the degree of approximation is determined by the same theory of probability. Savage also rejected the need for a dualistic view of probability in inference, which allows both objective and subjective ones.

Objectivity in science

No one will stop you from betting your own money, but how can we justify subjective beliefs in science? We may define objectivity as the agreement between reasonable minds, excluding the possibility of differing individual preferences. Savage put forward some arguments as to why there is no reason to exclude subjective probability as part of scientific reasoning. One argument is that, by consistency, two differing opinions will be brought closer by a sufficiently large sample size. The objective view presumes a common and universally accepted opinion as the goal of science, but in reality there are 'pairs of opinions, neither of which can be called extreme ... which cannot be expected to be brought into close agreement after the observational program' (Savage, 1972, page 68).

Ryder (1981) pointed out that an external agent can run a Dutch Book against two individuals that hold differing subjective probabilities of an event, i.e., make money from them regardless of what happens. Say John has a subjective probability 0.25 for re-election of Donald Trump, while Patrick has 0.15. Alice comes along to bet \$100 against John, and −\$100 against Patrick (meaning she plays the bookie against John, but lets Patrick be the bookie; or, even more clearly, she buys from Patrick and sells to John). If Trump is re-elected, Alice's gain is

$$g_1 = 0.25 \times 100 - 100 - 0.15 \times 100 + 100 = 10,$$

while, if he is not, she gains

$$g_2 = 0.25 \times 100 - 0.15 \times 100 = 10.$$

The only way to avoid the Dutch Book is for John and Patrick to have the same probability. From the financial-market perspective, the two players behave like two mini-markets that offer different prices, which provide an arbitrage opportunity – a risk-free investment – to an external agent. Note that both John and Patrick are individually coherent, and if there is no communication between them, they are none the wiser about the Dutch Book run against them. But if they are closely related individuals with regular contacts, and especially with a joint economy, we could say they are incoherent, because they have allowed a Dutch Book against them. A more detailed discussion was given previously in Section 6.6.

If the analogy is not too far-fetched, a collection of scientists in a scientific area is meant to be a closely communicating group with a common interest. Therefore, if individual scientists have differing subjective probabilities of relevant events, the group can be said to be incoherent. This might explain the reluctance of non-Bayesians or science in general to allow subjective preferences in the *formal assessment* of evidence.

Power and limit of axiomatic systems

An axiomatic system carries the strength of logic: Once we agree with the axioms, then we must agree with their implications. Even more than that, Savage viewed Bayesian statistics not simply as about attaching a prior distribution to a statistical problem, but as a holistic framework for making rational decisions under uncertainty. For him, the framework is strong enough to carry a normative force, not merely a tool to predict human behaviour.

However, in Chapter 3 we discussed that as the foundation of a mathematical structure, the axioms are true by definition, not to be proved or disproved. So, normative power of an axiomatic system does not come from the truth of its axioms. As assumptions for a model of reality, they can and should be checked empirically. Savage's view that his axiomatic system has the normative power of classical logic is not universally accepted. Shafer (1986) revisited Savage primarily to object to the normative aspect of the theory, in light of the accumulating empirical evidence that people violate the postulates. We will discuss paradoxes associated with Savage's axioms in detail in Chapter 14.

What prior?

For statistical applications, Savage never discussed the problem of how to choose a prior distribution. How do we specify an explicit set of preferences that satisfy both our personal wish and his axioms? This issue was already considered by many writers since the 19th century as a weakness in the inverse probability method. Perhaps, as a truly subjective Bayesian, he did not see the need to do that. In principle, any choice of a prior distribution will correspond to a set of preferences that satisfy his axioms, so there is no need for an explicit set of preferences. Much of the problem of prior selection is currently addressed by the so-called objective methods, such as Jeffreys's non-informative prior (Chapter 9).

8

The Inverse Probability Method

Classical writers such as Pascal and Leibniz used probability to quantify the uncertainty of events or propositions. The probability values were typically assumed, by appealing to simple arguments comparable to the principle of insufficient reason. This worked well for the games of chance. For other early applications, such as the analyses of birth or mortality tables, Bernoulli's and de Moivre's limit theorems allowed large-sample estimation of probability. But no exact method existed for estimating probability or updating it with data from small samples; recall Hume's struggle in Section 1.2. This is what's offered by the 'inverse probability' method, the dominant method of inference during the 19th century.

An exhaustive history of the method has been given by Dale (1999), so we shall focus only on Bayes and Laplace. Bayes's 1763 *Essay Towards Solving a Problem in the Doctrine of Chances* became the founding publication of the Bayesian School. Laplace's *Memoir on the Probability of Causes by Events*, published in 1774 when he was 25 years old, contained ground-breaking formulations of the Bayesian method; his *Analytical Theory of Probability*, first published in 1812, became the primary reference for probability for the rest of the century. Its trace was still visible in Fisher's first paper in 1912. Although Bayes's paper appeared earlier, it was clear that in 1774 Laplace had no knowledge of Bayes, whose *Essay* was largely ignored until the early 20th century.

The term 'inverse probability' itself did not appear in either Bayes's or Laplace's works. It appeared explicitly in the lecture notes by Fourier from the early 1800s and in de Morgan's (1838) *Essay on Probabilities.* Laplace referred to the probability of causes and future events given observed events, where 'cause' is the underlying parameter of a probability distribution. As the Anglo-Saxon countries became the centre of statistical development during the 20th century, the inverse probability method became known as the Bayesian rather than the Laplacian method.

8.1 Bayes's *Essay*

Bayes's *Essay* begins with a Problem section, containing just one exemplary statement (with italics as in the original):

> *Given* the number of times in which an unknown event has happened and failed: *Required* the chance that the probability of its happening in a single trial lies somewhere between any two degrees of probability that can be named.

In modern notation, the problem is as follows: given y successes in a binomial experiment with n trials and probability θ, what is

$$\Pr(\theta_1 < \theta < \theta_2 | y),$$

for any specified lower and upper bounds θ_1 and θ_2? He was using both the words 'chance' and 'probability' to help distinguish $\Pr(\cdot)$ from θ, but in his Section 1 he wrote that they mean the same thing. So what did he mean by the probability? If θ is a *fixed* unknown number, then the required chance is trivially either 0 or 1, which is not interesting. The answer to the problem appeared in his Proposition 10:

$$\Pr(\theta_1 < \theta < \theta_2 | y) = \frac{\int_{\theta_1}^{\theta_2} \theta^y (1-\theta)^{n-y} d\theta}{\int_0^1 \theta^y (1-\theta)^{n-y} d\theta}. \tag{8.1}$$

This is the standard Bayesian solution, with θ assumed to have uniform prior between 0 and 1. But Bayes justified this assumption by first proposing an auxiliary experiment involving a table ABCD 'so made and levelled' that if a ball W 'be thrown upon it, there shall be the same probability that it rests upon any one equal part of the plane as another.' (Just imagine a billiard table and ball, with ABCD marking the four corners of the table.) A line parallel to AD is then drawn from the resting position of the ball W; the line splits the table into two parts, say part-1 and part-2; let θ be the proportion of part-1. Then, a second ball is thrown n times, and y is the number of times it rests on part-1.

The essay then goes through a very lengthy and bewildering analysis of this ball experiment, which we can summarize very briefly: θ is random uniform on $[0, 1]$, and conditional on θ, the outcome y is binomial with parameter (n, θ). Under this construction, we get the conditional probability (8.1) as the answer to the Problem.

Fisher was very respectful of Bayes's genuine effort to justify his uniform prior distribution with the auxiliary experiment, suggesting that Bayes was not a real Bayesian in the modern sense. However, Bayes included a 'Scholium' in the *Essay*, where he wrote

> the same rule is the proper one to be used in the case of an event concerning the probability of which we absolutely know nothing antecedently to any trials made concerning it, ...

This sounds like Bernoulli's principle of insufficient reason. He continued: concerning such an event, it is 'as if its probability had been *at first unfixed* [our italic].' Unfixed by what, we may ask. Clearly, by ignorance. So, Bayes was indeed a Bayesian. If θ is not an outcome of a random experiment, the probability statement (8.1) is only meaningful within the Bayesian framework, where probability is used to represent ignorance and incomplete information.

To appreciate Bayes's seminal contribution, let's bring ourselves back to 1763. The title of Bayes's *Essay* is reminiscent of de Moivre's *Doctrine of Chances*, whose second edition in 1738 contains the proof of the normal approximation to the binomial distribution. In his cover letter, when submitting Bayes's *Essay* to the Royal Society, Bayes's close friend, Richard Price, indeed mentioned de Moivre – 'the great improver of this part of mathematics' – to motivate the novelty value of Bayes's result. Price first presented Bayes's Problem as the converse of de Moivre's. One could use de Moivre's result to estimate probability and set up an approximate normal-based confidence interval for the fixed true proportion. However, Price wrote, de Moivre's result depends on an 'infinite number of trials,' so its application is not exact, while Bayes's result is exact in small samples.

On the philosophical side, recall Hume's (1748) struggle with inductive reasoning (Section 1.2). As there was no rational/mathematical way to incorporate new information, Hume simply attributed inductive inference to a simplistic principle of 'Custom or Habit.' So Bayes's result could be seen as an answer to Hume's challenge.

What is Bayes's theorem?

The main result in Bayes's *Essay* is formula (8.1), but that's not what we call Bayes's theorem. Many texts, including Wikipedia, refer to this conditional probability as Bayes's theorem, formula or rule:

$$\Pr(B|A) = \frac{\Pr(A \cap B)}{\Pr(A)} = \frac{\Pr(A|B)\Pr(B)}{\Pr(A)}. \tag{8.2}$$

Why do we invoke Bayes, when we are just using Kolmogorov's definition of conditional probability? Moreover, the concept of conditional probability already appeared in de Moivre's classic book from the early 1700s. There is no good explanation. Bayes started his *Essay* with an operational definition of probability and its properties.

> **Definition:** The *probability of any event* is the ratio between the value at which an expectation depending on the happening of the event ought to be computed, and the value of the thing expected upon it's happening.

This is not the same as the classical definition of probability as the proportion of favourable outcomes. The conditional probability then appears in his Proposition 5:

> If there are two subsequent events, the probability of the 2nd is b, the probability of both together a, and the 2nd event has happened, "from hence I guess that the 1st event has also happened, the probability I am in the right" is a/b.

This corresponds exactly to formula (8.2), so perhaps it is not completely wrong to call it Bayes's rule.

According to Price, Bayes 'made an apology for the peculiar definition he has given of the word chance or probability. His design herein was to cut off all dispute about the meaning of the word, which in common language is used in different senses by persons of different opinions, and according as it is applied to past or future facts.' It is clear in the *Essay* that by 'expectation' he meant a wager. The 'value of the thing expected upon it's happening' is the amount received if the event happens. So, his probability is, in fact, a betting quotient, a precursor of Ramsey and de Finetti around 150 years later. However, Bayes's probability was not subjective in the sense that it could be specified via a coherency axiom a la Ramsey-de Finetti. Unfortunately, Bayes never solved other statistical problems, such as estimation of the general location parameter, which can be anywhere on the real line, so it is not clear if or how he would start with a physical experiment.

8.2 Bayesian Learning

Even though it's short, but because of its fundamental importance, we put the idea of Bayesian learning into its own section. Suppose that we do not know whether G is true or false, but we presume it has a nonzero probability $\Pr(G)$. Then, based upon the current evidence E, we can update our uncertainty about G via the Bayes rule

$$\Pr(G|E) = \frac{\Pr(E|G)\Pr(G)}{\Pr(E)}, \tag{8.3}$$

where $\Pr(E) = \Pr(E|G)\Pr(G) + \Pr(E|\text{not } G)\Pr(\text{not } G)$. The probability $\Pr(G)$ is the prior probability of G and $\Pr(G|E)$ the posterior. The probability $\Pr(E|G)$ is a special quantity called the likelihood of G, described in detail in Chapter 10. Bayes's formula describes how the prior information $\Pr(G)$ is updated to $\Pr(G|E)$, using new evidence E. Since $\Pr(E)$ is also a prior probability of E, the new information is captured in the likelihood. So we arrive at a general formula of Bayesian learning:

$$\text{Posterior} \propto \text{Likelihood} \times \text{Prior}. \tag{8.4}$$

As long as we have a prior $\Pr(G)$ and can compute the likelihoods $\Pr(E|G)$ and $\Pr(E|\text{not } G)$, we can compute the posterior $\Pr(G|E)$. This update of probability using Bayes's rule is key to recent algorithms used in machine learning and artificial intelligence.

8.3 Laplace's *Memoir*

Laplace made a careful distinction between two types of problems: prospective and inverse problems. Prospective problems refer to situations where the occurrence of an event is uncertain, but the underlying cause or factor is known. On the other hand, inverse problems arise when the event has already been observed, but the cause or factor behind it is unknown. In this context, the term 'cause' refers to the underlying parameter of the probability model being used, so the second type of problem is essentially an estimation problem. The *Memoir* is focused on this type, particularly on finding the unknown binomial probability and the probability of $X = x$ successes in m future trials, given $Y = y$ successes in n trials.

It is interesting to first see how the prediction problem motivates a Bayesian formulation: If the binomial probability θ is *fixed and known*, and X and Y are based on independent trials, then the conditional probability of X given $Y = y$ is free of y, so Y carries no information about X. But if θ is *unknown*, then logically and intuitively Y should carry information about X. But if we keep θ fixed, then the conditional distribution of X given $Y = y$ is still free of y, so perhaps surprisingly Y carries no information about X. 'Unfixing' θ – i.e., turning θ into a random variable – allows a non-trivial and informative conditional probability of X given $Y = y$. This is what Laplace did.

He started with a 'principle,' which in modern notation can be written as

$$\Pr(\theta_i|E) = \frac{\Pr(E|\theta_i)}{\sum_{j=1}^{n} \Pr(E|\theta_j)},$$

for a random event E and parameters $\theta_1, \ldots, \theta_n$. This is equivalent to Bayes's theorem, with the parameters $\theta_1, \ldots, \theta_n$ assumed to be complete (i.e., their probabilities add up to 1) and have uniform prior probabilities. Laplace made no distinction between discrete and continuous parameters, so for the binomial problem, the formula leads immediately to

$$\Pr(\theta|y) = \frac{\theta^y(1-\theta)^{n-y}}{\int_0^1 \theta^y(1-\theta)^{n-y}d\theta}, \tag{8.5}$$

and predictive probability

$$\Pr(X = x|y) = \binom{m}{x} \frac{\int_0^1 \theta^{x+y}(1-\theta)^{m+n-x-y}d\theta}{\int_0^1 \theta^y(1-\theta)^{n-y}d\theta}. \tag{8.6}$$

In particular, for $m = 1$ future trial,

$$\Pr(X = 1|y) = \frac{\int_0^1 \theta^{y+1}(1-\theta)^{n-y}d\theta}{\int_0^1 \theta^y(1-\theta)^{n-y}d\theta} = \frac{y+1}{n+2},$$

known as Laplace's rule of succession. So, Laplace came up with different results from Bayes, though, of course (8.5) is mathematically equivalent to Bayes's (8.1). The *Memoir* contained other mathematical results, including large-sample approximations of the posterior distribution and estimation of the location parameter.

8.4 Testimony Puzzle

Laplace was a great believer in the utility of probability theory for many areas of human activities. His *Philosophical Essay on Probabilities*, first published in 1814, contains 10 chapters on numerous applications, including the analyses of mortality tables, the mean duration of life and marriage, etc. Chapter XI discusses the probability of true testimony to assess the veracity of witnesses. We'll put the story in the form of a puzzle to highlight a surprising probabilistic phenomenon.

Suppose that witnesses are known to be truthful 90% of the time and to lie 10% of the time. (Deliberate lying is not necessary, in which case we can interpret the truthfulness probability as the reliability of witnesses.) In the following two situations, which witness report is more trustworthy?

A. A ticket is drawn at random from a box that contains 1000 tickets, 999 of which are black and 1 white. The witness reports that a white ticket is obtained (of course, without showing the ticket).
B. A ticket is drawn at random from a box that contains 1000 tickets, numbered $1, \ldots, 1000$. The witness reports that ticket number 79 is obtained (again without revealing the ticket).

The probability of drawing a white ticket in A is $1/1000$, so the report is rather suspicious, but the probability of ticket number 79 in B is the same. However, intuitively we know that the number could be anything, so it does not feel surprising or suspicious.

What's the effect of the possibility that the witness might lie? Were the witness 100% truthful, the report of a white ticket is still surprising, but the number 79 is not. So there is something fundamentally different about the two settings. For the coloured tickets, what we want is

$$\Pr(\text{True}|\text{White}) = \frac{\Pr(\text{White\&True})}{\Pr(\text{White})}.$$

By working out the following:

$$\begin{aligned}
\Pr(\text{White\&True}) &= \Pr(\text{White}|\text{True})\Pr(\text{True}) \\
&= 0.001 \times 0.9 = 0.0009, \\
\Pr(\text{White}) &= \Pr(\text{White}|\text{True})\Pr(\text{True}) + \Pr(\text{White}|\text{Lie})\Pr(\text{Lie}) \\
&= \Pr(\text{White}|\text{True})\Pr(\text{True}) + \Pr(\text{Black})\Pr(\text{Lie}) \\
&= 0.001 \times 0.9 + 0.999 \times 0.1 \\
&= 0.1008,
\end{aligned}$$

we get

$$\Pr(\text{True}|\text{White}) = \frac{9}{1009}.$$

Note the key step $\Pr(\text{White}|\text{Lie}) = \Pr(\text{Black})$, for to report white, the liar must have taken a black ticket. Therefore, the witness report of a white ticket is indeed suspicious. For ticket number 79, we can get similar probabilities

$$\begin{aligned}
\Pr(79\&\text{True}) &= \Pr(79|\text{True})\Pr(\text{True}) \\
&= 0.001 \times 0.9 = 0.0009, \\
\Pr(79) &= \Pr(79|\text{True})\Pr(\text{True}) + \Pr(79|\text{Lie})\Pr(\text{Lie}).
\end{aligned}$$

The tricky part is for the liar:

$$\begin{aligned}
\Pr(79|\text{Lie}) &= \Pr(\text{not getting 79 \& reporting 79}) \\
&= \frac{999}{1000} \times \frac{1}{999} = 0.001,
\end{aligned}$$

as the liar must have seen a number other than 79, then chose a number among 999 possible numbers. The probability $1/999$ presumes that the liar has no preference for any particular number. This means, interestingly, the reported number is independent of the trustworthiness of the witness. So, we have $\Pr(79) = 0.001$ and

$$\Pr(\text{True}|79) = 0.9,$$

meaning that the witness report is as trustworthy as you'd expect from an average witness.

What's the difference between the two settings? In the coloured ticket version, when faced with a black ticket, the liar can only choose white to report. But in the numbered version, the liar has 999 other tickets to choose from, thus bringing the probability of any chosen ticket to its original value.

The testimony problem is puzzling enough that John Stuart Mill in his 1843 book *A System of Logic Ratiocinative and Inductive* wrote that Laplace was mistaken, arguing that the two reports carry the same veracity. But, in the 1846 edition of the book, Mill conceded that Laplace was right after all. We can see why from Laplace's discussion of a scenario where the liar has a special interest in number 79. For instance, he might plan to say 79 with probability 1/2 or 1/3, etc. If he chooses 79 with probability 1, the probability $\Pr(\text{True}|79) = 9/1008$, the same as in the white ball case. But, of course, it's a very strong assumption to make about a generic liar we otherwise know nothing about.

8.5 The Sunrise Problem

The sunrise problem first appeared in Price's appendix to Bayes's *Essay*, where Price described several simple applications. If a man had been living in a cave and just came out and saw the sun for the first time, then saw it returning on the following day, what is his best guess the sun would return for the second time? Price actually gave odds of 3 to 1, but, counting the sun *returns* as the Bernoulli trials, according to Laplace's rule of succession with $n = 1$,

$$\Pr(X = 1|y = 1) = 2/3.$$

Price obviously meant to use Bayes's result (8.1), so to get a 3-to-1 odds he must have computed

$$\Pr(0.5 < \theta < 1|y = 1) = \frac{\int_{0.5}^{1} \theta d\theta}{\int_0^1 \theta d\theta} = 3/4,$$

thus interpreting the event of a sunrise tomorrow as having an underlying probability greater than 0.5, perhaps meaning the event is then no longer due to chance alone.

Assuming that we know no physics, the event E that 'the sun will rise tomorrow' continues to be governed by Laplace's rule. Assuming the young-earth creationist reading of the Bible, the number of days n can be estimated from the time the universe was created approximately 6000 years ago. So, given $n = 6000 \times 365.25 = 2,191,500$ days of consecutive sunrises, we get

$$\Pr(E|n \text{ consecutive sunrises}) = 0.9999995.$$

The probability eventually increases to 1 as n increases, but never reaches 1, which seems acceptable.

But, unfortunately, Laplace's formula is not good for assessing the proposition G that 'the sun always rises' or 'the sun rises forever,' which is equivalent to hypothesis $\theta = 1$. This is a Popperian scientific hypothesis, in the sense that it can be falsified by a conflicting observation, i.e., a day when the sun does not rise. We can view it as a model for the general truth of scientific laws or theories, such as Newton's laws or Einstein theories, that are supposed to hold indefinitely. Based on finite observations to date, is it possible to believe in $\theta = 1$ completely?

Given n days of consecutive sunrises, Laplace's formula (8.6) gives

$$\begin{aligned}&\Pr(n+m \text{ consecutive sunrises } |n \text{ consecutive sunrises})\\&=(n+1)/(n+m+1).\end{aligned}$$

Because 'the sun rises forever' means that m goes to infinity, it follows that

$$\Pr(G|n \text{ days of consecutive sunrises}) = 0 \text{ for all } n.$$

Hence, it seems we can't attach even a moderate probability to a general proposition where the possible or future tests of the rule are many times more numerous than the instances already investigated. The result implies that there is no way to have full confidence in general scientific theories based on finite data. It feels very disappointing, but it does correspond to Popper's dictum that only falsification by observation is possible in science. The problem here is that we already anticipate that the theory cannot possibly work over time, even though it has so far worked without fail.

Jaynes (2003) pointed out that a beta prior density, Beta(α, β) with $\alpha > 0$ and $\beta > 0$, can be interpreted as the state of knowledge that we have observed, a priori to the experiment, α successes (sunrises) and β failures (no sunrises). The uniform prior is Beta$(1, 1)$, which means that it assumes one day without sunrise a priori; this implies G can't be true or $\Pr(G) = 0$ a priori. In other words, no matter how long we have a consecutive record of sunrises, there is no way to confirm the proposition G. The uniform prior is meant to represent ignorance. But this is actually not the case, as it leads to a very strong presumption that $\Pr(G) = 0$ and, consequently,

$$\Pr(G|E) = \Pr(G) = 0.$$

This specific issue has a more general relevance: In his *Logic of Scientific Discovery*, Popper (1959, Appendix vii) explains why he believes $\Pr(G) = 0$. His idea is such that no evidence can alter the probability of the general proposition $\Pr(G|E) = 0$, thus ruling out probability-based induction. We discuss this further in the next section.

Jeffreys's resolution

The reason that a general proposition cannot be confirmed by scientific induction, as seen in Laplace's formula, turned out to be because of the wrong choice of the prior. Jeffreys's (1939) resolution to the problem was to find another prior, which places a point mass 1/2 on the general proposition G: $\theta = 1$ and a uniform prior on [0,1] with 1/2 weight. Let E be an event of n days of consecutive sunrises. This leads to

$$\begin{aligned} 0 &< \Pr(G) = 1/2 \\ &< \Pr(G|\text{one day of sunrises}) = 2/3 \\ &< \Pr(G|\text{two days of consecutive sunrises}) = 3/4 \cdots \\ &< \Pr(G|E) = (n+1)/(n+2) \cdots \end{aligned}$$

And thus, $\Pr(G|E)$ increases to 1 eventually, as n increases (Lee, 2020).

A key to Jeffreys's resolution is to presume a priori that $\Pr(G) = 1/2$. In general, for those who believe $0 < \Pr(G) < 1$, evidence can update their probability. Jeffreys's resolution led to an important innovation of the Bayesian school, bringing 'a touch of genius, necessary to rescue the Laplacian formulation of induction' (Senn, 2009). However, with Jeffreys's resolution, scientific induction still cannot attain complete confidence, because such a process requires infinite evidence to get $\Pr(G|E) = 1$.

8.6 Qualitative Plausible Reasoning

The Bayesian formulation generates fruitful qualitative reasoning as an extension of classical logic. If G implies E and we observe E, what can we say about G? The classical syllogism is silent here. Bayes's formula (8.3) is normally used to compute numerical conditional probability, especially if both G and E are random events. However, it is instructive to see its qualitative use for logical scientific reasoning where G and E are propositions. We also see parallels with Polya's plausible reasoning (Polya, 1954), which he used to find support for interesting theorems in mathematics before they are proved. We can also use the formulation to justify and clarify abductive reasoning (Section 2.4).

In principle, if we subscribe to logical probability theory (Section 6.2), G can be any proposition: theorems in mathematics, basic propositions in philosophy, or any other proposition, such as the re-election of the current president. To illustrate briefly and qualitatively, let G be a general proposition, such as 'All ravens are black' or 'All men are mortal.' And let E be a particular proposition, an observation or a piece evidence that is implied or completely predicted by G; for instance, 'The raven in front of me is black,' or 'A man named Socrates died.' Then, representing deductive logic $G \to E$ in a probability language, we have

$$\Pr(E|G) = 1.$$

Using Bayes's formula (8.3), after observing the predicted E, thus implying that $\Pr(E) > 0$, we can update the degree of belief in G as follows:

$$\Pr(G|E) \quad = \quad \Pr(G)/\Pr(E) \geq \Pr(G),$$

with strict inequality if $\Pr(E) < 1$. Therefore, a particular observation E can corroborate or increase support for the general proposition G.

Clearly, if $\Pr(E)$ is small, observing the evidence would generate a strong increase in the belief of G. This is not intuitive. Typically, a good general theory G will produce multiple predictions. When planning a study to test G, an investigator should choose among competing experiments the one that would produce a priori *the least probable observation.* This useful notion is not obvious within the context of abductive reasoning.

For instance, the general theory of relativity predicts the bending of light by gravity, which was confirmed by measurements during the 1919 solar eclipse. Without Einstein's theory, who would have accepted that a light passage is bent by gravity or that space is curved? As another example, according to the standard theory in particle physics, we have no reason to suspect that protons would decay. Yet, this is what's predicted by the Grand Unified Theory (GUT). So, observing proton decay would provide very strong support for the GUT.

What if G does not imply E fully? Suppose E is more likely when G is true than when G is not true, i.e.,

$$\frac{\Pr(E|G)}{\Pr(E|\text{not } G)} > 1.$$

Then,

$$\begin{aligned}\frac{\Pr(G|E)}{\Pr(\text{not } G|E)} &= \frac{\Pr(E|G)}{\Pr(E|\text{not } G)} \times \frac{\Pr(G)}{\Pr(\text{not } G)}\\ &> \frac{\Pr(G)}{\Pr(\text{not } G)}.\end{aligned}$$

Thus, evidence E can enhance the belief in G. It is also clear that if the probability $\Pr(E|\text{not } G)$ is much smaller than $\Pr(E|G)$, then E can provide very strong support for G. This suggests that, as a good test of G, we need to perform an experiment to produce evidence E that is likely to occur if G is true, but highly unlikely if G false. Furthermore, the formula also explains Popper's non-intuitive suggestion that, among competing hypotheses, one should choose the least probable hypothesis that is supported by data, i.e., the posterior ratio $\Pr(G|E)/\Pr(\text{not } G|E)$ is high, while the prior ratio $\Pr(G)/\Pr(\text{not } G)$ is the smallest.

Imagine a detective interviewing crime suspects, where G means a suspect is guilty. Most suspects are innocent, so a priori, $\Pr(G)$ is small and $\Pr(\text{not } G)$ large. The detective would be watching for behaviour patterns – including polygraph tests – that would be unusual if the person was not guilty but expected for a guilty person, i.e., $\Pr(E|\text{not } G)$ is low but $\Pr(E|G)$ is large.

Special case when $\mathbf{Pr}(G) = 0$ or $\mathbf{Pr}(G) = 1$

Assuming that E is observable such that $P(E) > 0$, if we start with $\Pr(G) = 0$ then $\Pr(G \cap E) = 0$, so from Bayes's formula we have

$$\Pr(G|E) = \Pr(G \cap E)/\Pr(E) = 0.$$

On the other hand, if $\Pr(G) = 1$ then $\Pr(G \cap E) = \Pr(E)$, so

$$\Pr(G|E) = \Pr(E)/\Pr(E) = 1.$$

Therefore, if you start with $\Pr(G) = 0$ or $\Pr(G) = 1$, no evidence E can change your belief in G. Kant (1781) simply claimed that certain propositions, such as Euclid's axioms, are valid a priori, i.e., $\Pr(G) = 1$. Actually, on the spherical surface of the earth, the parallel postulate G is false, but by starting with $\Pr(G) = 1$, no evidence could alter his posterior probability $\Pr(G|E) = 1$.

As Max Planck said, 'a new scientific truth does not triumph by convincing its opponents and making them see the light, but rather because its opponents eventually die, and a new generation grows up that is familiar with it.' Although Columbus presented convincing evidence of the spherical shape of the Earth, this did not change the belief of those who firmly believed that the Earth is flat.

At the other extreme, Popper (1959, Appendix vii) argued for $\Pr(G) = 0$, only to have $\Pr(G|E) = 0$. Therefore, inductive reasoning is impossible, and falsification is the only way to conduct scientific inference. Under $\Pr(G) = 0$ a priori, a universal law cannot be accepted but is also not rejected until conflicting evidence appears. As the philosopher Charlie Broad stated, 'induction is the glory of science, but the scandal of philosophy.'

Degree of confirmation

Carnap's (1950) degree of confirmation of general propositions G by evidence E is defined by

$$C(G, E) \equiv \Pr(G|E) - \Pr(G)$$

Popper (1959) preferred 'corroboration' over 'confirmation', and 'testability' over 'confirmability,' since a theory is never confirmed to be true, as for him $\Pr(G) = 0$. However, the term 'confirmation' has survived in the literature, so we also use it here to mean 'corroboration.' If we start with $\Pr(G) = 0$, we get $\Pr(G|E) = 0$ and $C(G, E) = 0$. Therefore, in Carnap's inductive logic, the degree of confirmation of all universal laws with presumed zero prior is always zero.

In Jeffreys's (1939) resolution of the sunrise problem, with $\Pr(G) = 1/2$,

$$C(G, E) = \Pr(G|E) - \Pr(G) = \frac{n}{2(n+2)} > 0,$$

and thus the evidence E supports the general theory G positively. Therefore, the choice of prior determines the degree of confirmation.

9

What Prior?

The Bayesian school stands on two main pillars: subjective probability and priors. The former is justified by Ramsey-de Finetti's coherency axiom (Section 6.5), which most people would find reasonable or even attractive, or Savage's axioms. Bayesians such as David Lindley like to put out this challenge: The Bayesian philosophy is axiomatic, so if you choose to be non-Bayesian, please explain what you find unacceptable in the axioms. However, even if we accept subjective probability, the question of what prior distribution to use in a specific analysis remains a stumbling block for many non-Bayesians. A subjective choice, which is justified within the subjective probability framework, implies arbitrariness.

Fisher was unequivocal in his objection to the Bayesian prior: Why include in the analysis an element that we cannot justify objectively? Full objectivity might be hard to achieve, since any experiment or data analysis involves making inevitable choices, but as in Keynes's argument for logical probability, 'objectivity' should at least mean an agreement between rational thinking individuals. These individuals might be the collection of scientists in a research area, if not the whole of science. Subjectivism in the choice of prior, leading to individualistic results, fails this objectivity test. In other words, *you can be coherent without being objective.* That is an answer to Lindley's challenge. From now on, we shall only discuss non-subjective non-informative priors. But here also there are specific objections due to paradoxes that arise from seemingly reasonable priors such as the uniform prior.

Not surprisingly, Bayesians have extensively investigated the problem of prior selection. Different variants of Bayesianism are identified not from the way they interpret probability – which is invariably of the epistemic kind – but from the way they choose a prior. The so-called objective Bayesians choose a non-subjective non-informative prior that produces a posterior with reasonable objective probability properties, thus mimicking the frequentist procedure. The term 'objective Bayesian' clearly indicates that being 'Bayesian' is not synonymous with being subjective. Still, the term is rather confusing since, presumably, objective Bayesians believe in subjective probability, just not in subjective priors.

Throughout the 19th century, the uniform prior was practically the only non-subjective prior used in the inverse probability method. By the end of the 19th century, many people realized the problems associated with the uniform prior. Modern theories of priors came only after the 1940s, particularly with Jeffreys's work on invariant priors and the development of objective and reference priors. There is also a philosophical quandary: The objective priors are often improper, i.e., not a valid probability. When improper, according to Ramsey-deFinetti's or Savage's fundamental theorems (Sections 6.5 and 7.4), the prior will lead to at least one violation of their rationality axioms. Thus, in general, objective Bayesians do not follow the standard Bayesian axioms faithfully, which in turn raises the question of how they justify their subjective probability.

The growth of non-Bayesian methods in the early 20th century, particularly those developed or originated by Fisher, brought the decline of the inverse probability method. For the rest of the century, the Bayesian school remained in the minority among statisticians, and Bayesian methods and quantities were rarely used by applied scientists. Most of the realistic data analysis problems solved by non-Bayesian methods were initially computationally too hard in their Bayesian version. However, the rise of computing power and computational algorithms – such as the Markov Chain Monte Carlo method to compute the posterior – has made the Bayesian methods attractive again, to the point that some statisticians become Bayesian in only methodology and computation, but not in philosophy.

9.1 The Principle of Insufficient Reason: Uniform Prior

In the classical era, Bernoulli's principle of insufficient reason was key for generating probabilities of random events; the probabilities were needed in games of chance or other simple applications. When the principle is based on the natural symmetry of the experiment, such as the similar sides of a coin or a die, it often leads to acceptable probabilities. In the

Bayesian methodology, the principle is used to justify non-informative prior distributions, which should represent ignorance of underlying parameters rather than random events.

The uniform prior – also known as the flat or vague prior – was the first to arrive on the scene. Laplace used it by default in his 1774 *Memoir* and again in his 1812 book. Bayes, however, justified the uniform prior by first imagining a physical experiment of rolling a ball on a perfectly flat table.

In Section 6.1 we discussed Bertrand's and von Mises's paradoxes, which arise due to non-unique solutions when the uniform distribution is used to represent the uncertainty of random outcomes. Here, we discuss the problems that arise when we use the uniform as a prior distribution for underlying parameters. Let's start with the Bernoulli trials considered by Bayes and Laplace, where we know nothing about the event. With a single trial $n = 1$, what is $\Pr(Y = 1)$? According to the principle of insufficient reason applied to random events, we should have $\Pr(Y = 1) = 1/2$. This matches the Bayesian solution assuming uniform prior on the parameter:

$$\begin{aligned}\Pr(Y = 1) &= \int_0^1 \Pr(Y = 1|\theta)d\theta \\ &= \int_0^1 \theta d\theta = 1/2.\end{aligned}$$

Now consider $n = 2$ independent trials: We can report the random outcome either in full as $x \in \{00, 01, 10, 11\}$ or as the total number of successes $y \in \{0, 1, 2\}$. Applying the principle, we assign a uniform probability of $1/4$ to x, but $1/3$ to y. For instance, $\Pr(X = 00) = 1/4$ and $\Pr(Y = 0) = 1/3$, but this is of course absurd because they are probabilities of the same random outcome. Which is more reasonable? If your immediate inclination points to the first assignment, think again.

Applying the principle of insufficient reason to the unknown probability θ, we actually have the usual Bayesian computation for the marginal probability of Y:

$$\Pr(Y = y) = \binom{2}{y} \int_0^1 \theta^y (1-\theta)^{2-y} d\theta = 1/3, \text{ for } y = 0, 1, 2,$$

which matches the second assignment! However, for X we have

$$\begin{aligned}
\Pr(X = 11) &= \int_0^1 \theta^2 d\theta = 1/3 \\
\Pr(X = 00) &= \int_0^1 (1-\theta)^2 d\theta = 1/3 \\
\Pr(X = 01) &= \Pr(X = 10) = \int_0^1 \theta(1-\theta) d\theta = 1/6.
\end{aligned}$$

Now the four random pairs have no uniform probability! The calculation is saying that if you know nothing about the probability of an event, then it is better, for instance, to bet on 11 than on 10. If $\theta < 0.5$ then $\theta^2 < \theta \times (1-\theta)$, and vice versa for $\theta > 0.5$, so the optimal bet is not so obvious. Yet, you won't come back to an unknown restaurant even after just one bad meal there. That's a bet on 11 over 10.

Thus the application of the principle of insufficient reason can result in a mismatch when applied to the outcome or to the underlying probability. What is going on here? Perhaps not immediately obvious, the assumed uniform θ induces a positive correlation between the first and second trial, such that, say, the probability of concordant outcomes 11 is higher than the probability of discordant outcomes 10. But, haven't we assumed the trials to be independent? (Technically, the analysis does assume conditional independence given θ, which induces marginal positive correlation when θ is 'unfixed.') Since the trial outcomes are observable, it is legitimate to ask where/how we can see the evidence of unequal probabilities and positive dependence. Overall, are these highly specific properties acceptable, considering that we are supposed to know nothing about the events?

On the other hand, assuming that the probability of each random pair is 1/4 implies that the two trials are independent, each with probability 1/2. But, aren't we supposed to be ignorant about the probability? Moreover, it precludes the possibility of learning: The two trials are now independent, so seeing 1 in the first trial does not tell us anything about the second trial.

From a statistical perspective, the issue arises because the principle of insufficient reason is taken as a guiding principle rather than a part of checkable modelling assumptions. Thus, a discrepancy in the results from two different ways of applying the principle looks confusing or even disturbing. Furthermore, the statistical consequences might only be implicit, so they are not immediately visible for checking. From the statistical modelling perspective, it is, of course, not surprising that two different assumptions lead to two different results, and as much as possible we *use the data* to judge which assumption is more reasonable.

In the binomial experiment, if justified, we can make an *explicit assumption* that θ is a random sample from uniform[0,1], *and* conditional on θ, the outcome y is a series of n independent Bernoulli trials. There is then no confusion between the two probability results. In a real data analysis situation, one may check, for instance, if the assumption of conditional independence is reasonable. We normally try to avoid assumptions that lead to uncheckable features, such as positive dependence in a single pair of Bernoulli trials. But if we have many observed pairs or a long time series, then dependencies can be checked.

An obvious and well-known general problem with the uniform prior is the lack of invariance. Assuming that we know nothing about a probability θ, the uniform[0,1] may seem sensible as a representation of ignorance. But if we are interested in another parameter $\phi = \theta^2$, the distribution of ϕ is still between 0 and 1, but it's no longer uniform: It's much more likely to be near zero, e.g., 50% of its probability is on $\phi < 0.25$. So, ignorance on one scale becomes informative on another scale, but we have no general reason to say that one scale is more natural than any other. For instance, the probability itself may look like a natural scale, but gamblers often think in terms of the odds $\theta/(1-\theta)$. Comparisons between probabilities may be more natural in terms of ratios, which suggest a log-scale, etc.

Still another problem arises when the parameter space is unbounded, so the uniform prior is improper with infinite total measure. The Bayesian update formula, Posterior $\propto$ Likelihood $\times$ Prior, is still assumed, but there is no guarantee that the posterior is proper and sensible; this requires a case-by-case verification. For instance, consider samples from $N(\mu, \sigma^2)$. For the mean μ, the natural parameter space is the whole real line, while for σ, the natural parameter space is the non-negative side, so uniform priors for μ and σ are both improper. It turns out, as we discuss next, that the uniform prior for μ is sensible, i.e., it leads to a sensible posterior, but not for σ.

9.2 Axiomatic Basis of the Principle of Insufficient Reason

As originally proposed by Bernoulli and Laplace, the principle of insufficient reason was justified on intuitive grounds. So paradoxes of the principle suggest only that sometimes our intuition fails, which is an acceptable event. However, Sinn (1980) showed that the principle can be justified axiomatically using two axioms from the rational choice theory a la von Neumann and Morgenstern or Savage. Hence, the principle has normative power if we accept the axioms.

Suppose that one is betting on a lottery or a horse race but completely ignorant about the probabilities. What is the rational choice for the probabilities? The theory allows both objective or subjective probabilities, which correspond to the lottery or the horse race respectively, but for linguistic convenience let's stick to the lottery. The payoff is one of $R_1 \ldots, R_m$, associated with the potential winning number among $s_1, \ldots, s_m$. The first axiom on weak ordering specifies:

> **Axiom 1**: There exists a complete and transitive weak ordering of the results $R_1 \ldots, R_m$.

This axiom corresponds to VNM's first and second axioms for their utility theory, and Savage's first axiom (Section 7.1). Sinn's second axiom is the axiom of independence, which roughly specifies the following:

> **Axiom 2**: The ordering of two lotteries is independent of results that are the same for the two lotteries.

This axiom corresponds to VNM's independence axiom and Savage's sure-thing principle. A formal statement of the axiom and further discussion can be found in Axiom 2 in Section 7.1. Sinn (1980) proved the following:

> **Theorem**: Under complete ignorance and the two axioms, bets must be evaluated as if $s_1, \ldots, s_m$ have equal objective probabilities.

This indeed corresponds to the principle of insufficient reason. Unfortunately, the proof depends crucially on the discrete outcomes, so it does not solve the paradoxes involving continuous outcomes. Moreover, as we discuss in Chapter 14, the sure-thing principle is not uncontroversial and there are paradoxes associated with it. If the outcome is the whole real line, the principle leads to an improper prior, so according to Savage's fundamental theorem (Section 7.4), it will violate at least one of his rationality axioms. So, a general axiomatic support for the principle of insufficient reason must be found using other ways.

9.3 Invariant Prior*

The uniform prior ruled the inverse probability method for at least 150 years. The lack of invariance with respect to the choice of scale means that any specific choice of scale is arbitrary, but it was not clear that there was a good way out. In his classic book *The Laws of Thoughts*, Boole (1854, p. 384) wrote:

> [such arbitrariness] seems to imply that a definite solution is impossible, and to mark the point where inquiry ought to stop.

The first and perhaps the most influential effort to get a non-uniform non-informative prior was by Harold Jeffreys – a statistician, geophysicist and astronomer, and a contemporary of Fisher in Cambridge – particularly in his classic *Theory of Probability*, first published in 1939. To a large extent Jeffreys's goal was an objective Bayesianism that achieved Keynes's idea of logical probability as a degree of rational belief. It's a belief agreed upon by rational thinking individuals, much like they agree on classical deductive logic; see Section 6.2. (Jeffreys made it clear, however, that he was not a follower of Keynes! It just happened that they both attended the same lectures by W.E. Johnson. There is one big difference in their views of probability: Keynes's probability is partially ordered, whereas Jeffreys's probability axioms specify complete ordering.)

The invariance requirement is simple enough: if θ is assumed to have density $f(\theta)$, and $\phi \equiv h(\theta)$ is a 1-1 function of θ, then the prior of ϕ must satisfy the relationship between probability densities:

$$f(\phi) = f(\theta)|J|, \tag{9.1}$$

where $|J| = |\partial\theta/\partial\phi|$ is the Jacobian term. This means that we get consistent inferences regardless of how we choose to parameterize the problem. For discrete θ, invariance automatically holds because there is no need for a Jacobian term. Therefore, the invariance issue arises only with continuous parameters.

For instance, consider the estimation of the standard deviation σ. We have already discussed that the uniform prior lacks invariance: If σ is uniform and $\phi = \sigma^k$, then $|J| = \sigma/(k\phi)$, so $f(\phi)$ is not uniform. Now consider the prior of the form $f(\sigma) \propto 1/\sigma$. Then

$$f(\phi) \propto 1/\sigma \times \sigma/\phi = 1/\phi,$$

so the prior is invariant for any k in the class of functions σ^k with $k \neq 0$. Alternatively, we can obtain this prior by setting $\log\sigma$ to be uniform. Although invariance is not guaranteed for all 1-1 functions, the prior $1/\sigma$ is known as the invariant prior for σ.

The same reasoning applies to any positive parameter. For example, for the Poisson mean parameter θ, we might consider $1/\theta$ as the invariant prior. What about the binomial probability θ? We can first transform it to $\phi \equiv \theta/(1-\theta)$, which now covers the positive real line, and then assume that $\log\phi$ is uniform. This implies $f(\theta) \propto 1/\{\theta(1-\theta)\}$, which is also known as Haldane's prior.

But how do we get a prior that is invariant across the wider class of all 1-1 functions? This is achieved by the so-called Jeffreys's prior (now without the word 'invariant'):

$$f(\theta) \propto |\mathcal{I}(\theta)|^{1/2}, \tag{9.2}$$

where $\mathcal{I}(\theta)$ is the expected Fisher information

$$\mathcal{I}(\theta) = -E\frac{\partial^2 \log L(\theta; Y)}{\partial\theta^2},$$

and Y is the outcome variable. The prior $f(\theta)$ is then invariant across all 1-1 functions, since the necessary Jacobian term for the transformation also appears in the expected Fisher information. In the multiparameter case, formula (9.2) is interpreted as the square-root of the determinant. To give some well-known examples:

(i) For the normal case $N(\mu, \sigma^2)$ we get sensible results: if σ^2 is known, Jeffreys's prior for μ is constant. And if μ is known, the prior for σ is $1/\sigma$, which matches the previous invariant prior.

(ii) For the Poisson mean case, we get the prior $f(\theta) \propto 1/\sqrt{\theta}$, which is in conflict with the previous invariant prior. Jeffreys (1961, p. 186) argued that if we have 'complete ignorance' about θ, then $1/\theta$ is more appropriate. But if we're going to perform a Poisson experiment, we will likely set it up such that θ is of moderate value, not too small; in this case $1/\sqrt{\theta}$ is more appropriate. This illustrates that when using Jeffreys's prior, we are already using some features of the data. It also shows that the 'invariance' requirement is not unique.

(iii) For the binomial experiment with probability θ, we get the prior $f(\theta) \propto 1/\sqrt{\theta(1-\theta)}$, compared to Haldane's prior $f(\theta) \propto 1/\{\theta(1-\theta)\}$.

Jeffreys's prior has been justified from many angles, but perhaps the most attractive justification is its objective properties: In large samples, for single parameters, 95% credible intervals derived from the posterior will have approximately 95% long-term frequentist probability (Welch and Peers, 1963). This implies that the probability derived from the Bayesian posterior can be interpreted as Keynes's logical probability, which is agreeable to rational thinking individuals.

As a final comment, Jeffreys's prior is invariant with respect to a wide class of transformations at the expense of having to consider the probability model for the data Y. Philosophically, this is rather problematic, since a prior is supposed to be completely free from the data or the sample space of the experiment. For instance, suppose y is binomial (n, θ), then Jeffreys's prior is $1/\sqrt{\theta(1-\theta)}$. But if y is negative binomial, the prior is $1/\{\theta\sqrt{1-\theta}\}$. Should the prior, which supposedly represents ignorance, be affected by how we plan our experiment, or worse, by our intention on how we model the data?

9.4 Multiparameter Case*

There was an initial optimism that Jeffreys's prior would solve all the prior selection problems. It works well for single parameters, but difficulties arise when dealing with multiple parameters. They already appear in the normal case: When the mean and variance parameters are both unknown, Jeffreys's joint prior is

$$f(\mu,\sigma) \propto |\mathcal{I}(\mu,\sigma))|^{1/2} = 1/\sigma^2,$$

but, intuitively, ignorance of μ should be independent of σ, so their priors should independent, i.e.,

$$f(\mu,\sigma) \propto 1/\sigma.$$

The latter is indeed better than the former. How can we judge that? By comparing the resulting posteriors of σ: the former leads to

$$\sum_i (y_i - \bar{y})^2/\sigma^2 \sim \chi^2_n,$$

which shows an incorrect degree of freedom for the χ^2 distribution, but the latter leads to the correct degree of freedom (Bernardo and Smith, 1994). This raises one salient point: *it is sensible to judge a prior* by looking at the resulting posterior.

Another serious issue is *the so-called marginalization paradoxes*, an indication that the posterior is not a proper probability distribution. The simplest one, first found by Stone and Dawid (1972), involves finding the posterior of $\phi = \mu/\sigma$ in the normal model. They showed that, starting with the joint prior $f(\mu,\sigma) \equiv 1/\sigma$, the marginal posterior of ϕ depends only on a statistic $t \equiv n\bar{y}/\sqrt{\sum y_i^2}$. Now the distribution of t depends only on ϕ, so we can also get the posterior $f(\phi|t)$ directly using Bayes's update formula

$$f(\phi|t) \propto f(t|\phi)f(\phi).$$

But there is no prior $f(\phi)$ that would match the two versions of the posterior. This marginalization problem turns out to be common in multiparameter problems (Dawid et al., 1973).

All these problems and the current resolutions indicate that the notion of complete ignorance is too vague to be represented probabilistically. The data, the experiment and, most importantly, the resulting posterior also play a role in the prior selection. The best one can hope for seems to be that the prior should not be influential relative to the data in determining the posterior, or vice versa, that the data must somehow be dominant. Philosophically, this is an objectivist or empirical attitude as opposed to the normative attitude of de Finetti's or Savage's.

Stein's paradox (Stein, 1959; Bernardo and Smith, 1994) illustrates a case where the prior dominates the data. Let $y_1, \ldots, y_n$ be iid random vectors from $N(\mu, I_p)$, where $y_i \equiv (y_{i1} \ldots y_{ip}) \in R^p$ and $\mu \equiv (\mu_1 \ldots \mu_p) \in R^p$, and I_p is a $p \times p$ identity matrix. That is, y_i is sampled from independent normals with different means. The parameter of interest is

$$\theta = \sum_{j=1}^{p} \mu_j^2.$$

Since μ is a location parameter, one might consider the non-informative prior $f(\mu) \propto 1$. Setting $t \equiv \sum_{j=1}^{p} \bar{y}_j^2$, we can show that the posterior mean is

$$E(\theta|y) = t + p/n,$$

which typically can be used as an estimate of θ. But, based on $E(t|\theta) = \theta + p/n$, the method-of-moment estimate is

$$\widehat{\theta} = t - p/n.$$

When p is large relative to n, there is a large discrepancy between $\widehat{\theta}$ and the posterior mean. Here, the method-of-moment estimate is closer to the data, but the posterior estimate is primarily driven by the prior being too diffuse.

In general, a prior and its posterior distributions will be different, so there is a distance between them. Different priors will intuitively lead to different distances. The so-called reference prior (Bernardo, 1979) is the prior that has the maximum possible distance. In a sense, it gives the maximum possible leverage to the data. In the single-parameter case, it turns out to be equivalent to Jeffreys's prior, but in the multiparameter case it can deal properly with the marginalization and Stein's paradoxes. Progress in the general method for dealing with multiparameter priors appears to rely on the reference prior, but the details are too technical for this book.

Aerospace scientists recognize a probability dilution puzzle where the probability of satellite collision decreases as the uncertainty in satellite trajectories increases. Since trajectory uncertainty is driven by errors in the tracking data, the absurd implication of probability dilution is that lower-quality data reduce the risk of collision. This so-called satellite conjunction paradox can be represented by a two-dimensional model (Balch et al., 2019) such that $y_1 \sim N(\mu_1, \sigma^2)$ and $y_2 \sim N(\mu_2, \sigma^2)$ represent the coordinates and θ the squared distance of the two satellites.

This can be viewed as Stein's paradox with $p = 2$ and $n = 1$. In the satellite conjunction paradox, the probability dilution becomes apparent as $\sigma \to \infty$. In Stein's paradox, the probability dilution becomes severe as $p \to \infty$. In both of these paradoxes, the use of a reference prior improves the uniform prior to moderate the probability dilution effect. However, such a probability dilution can be avoided by using the confidence distribution in Section 11.3.

Part III

Likelihood and Likelihood-Based Inference

10

Likelihood

Among the technically distinct concepts of uncertainty, we have probability, confidence and likelihood. We can summarize briefly the differences as they apply to statistics. All orthodox statistical inferences are probability-based. Confidence and likelihood are derived from probability, but they are not probability, since they do not satisfy the probability axioms. The confidence concept is traditionally meaningful only within the theory of confidence intervals, so it seems to have limited reach as a measure of uncertainty. Likelihood is the central concept in statistical modelling and inference. Its role in modelling is universally recognized and accepted, but direct likelihood inference is controversial. Traditional likelihood-based inference requires probability-based calibration. The fundamental value of likelihood is theoretically established by the likelihood principle theorem, which implies that the evidence in the data is embodied in the likelihood function.

10.1 Classical Likelihood

As a 22-year-old undergraduate student in Cambridge, Fisher (1912) wrote the paper *On an Absolute Criterion for Fitting Frequency Curves.* The 'absolute criterion' is none other than the likelihood, although he did not use the word yet. It is 'absolute' in contrast to the arbitrariness of the method of moments or the method of least squares. However, indicating the then still influential Laplacian inverse probability method, he called the resulting estimate 'the most probable value.' The word 'likelihood' as a technical term appeared for the first time later in his 1921 paper. He contrasted it with 'probability' and emphasized that they are two 'radically distinct concepts [that] have been confused under the name of "probability"... .'

Through the method of maximum likelihood (Fisher, 1922), which is freed from the prior distribution, the likelihood became the central theoretical concept and tool that drove statistical modelling and inference, especially for the last five decades. It becomes indispensable as statistical applications move from Gaussian data with straightforward collection to non-Gaussian data or data with complex ascertainment, such as censored or incomplete data (Pawitan, 2004).

Formally, assuming the data y are generated from a statistical model parameterized by a fixed and unknown θ, we have the following:

> **Definition:** The likelihood $L(\theta; y)$ is the probability of the observed data y seen as a function of θ.

When there is no risk of confusion, we will drop the source data y and only write $L(\theta)$. The parameter θ and the data y can, in principle, be of arbitrary complexity. In some applications, θ may represent a simple hypothesis or proposition. For instance, in a criminal court, θ_0 is the hypothesis that the defendant is innocent, and θ_1 the defendant is guilty. In meteorology, θ could be a dynamic rainfall map of a geographical area over time.

Likelihood provides an immediate objective tool to compare hypotheses: θ_1 is preferred over θ_0 if $L(\theta_1) > L(\theta_0)$. If there is a range of θs under consideration, then we can simply search for the value that maximizes the likelihood, the so-called maximum likelihood estimate (MLE). This is the standard route to take for most applications of the likelihood. However, orthodox likelihood-based inference by Bayesian and frequentist schools still needs probability-based calibration. Edwards (1972) and Royall (1997) are rare proponents of using likelihood without probability as a tool to assess evidence. In his last book on statistical inference, Fisher (1959) suggested that direct likelihood inference be used in cases where the evidence is too weak to provide probability-based inference; he gave some discrete data examples as illustrations. However, in general, direct use of likelihood for inference is controversial and not widely accepted. There are many examples (Pawitan, 2001, Chapter 7), including the ticket paradox below, showing how the likelihood value alone can be misleading.

One of Fisher's great realizations, already clear from his 1912 paper, was that the likelihood is not a probability. In retrospect, this is quite easy to establish by checking the probability axioms, but those axioms only appeared explicitly in 1933. Being non-negative, the likelihood trivially satisfies the first axiom of probability. The second axiom is not satisfied because the likelihood is only defined up to a constant term relative to θ, so it does not add up or integrate to one. To see this, assume y from a continuous model $p_\theta(y)$. Formally, the probability of a continuous outcome y is zero, but in reality, y will always be measured with finite precision ϵ, so the likelihood is

$$L(\theta; y) = p_\theta(y)\epsilon.$$

Clearly, if we transform the data y with any 1-1 function $f(\cdot)$ to get $u \equiv f(y)$, we'd have the same amount of information on θ. But now the likelihood based on u is

$$L(\theta; u) = p_\theta(y)|J|\epsilon \propto L(\theta; y),$$

where $|J|$ is the Jacobian term, which is free of θ. Therefore, as a function of θ, the likelihood can vary up to a scale term without changing its information content. This is, in fact, the way Fisher justified in 1912 that his absolute criterion is not a probability. Proportional likelihoods are said to be equivalent, denoted by $L_1 \sim L_2$ to mean

$$L_1(\theta) = cL_2(\theta), \quad c > 0,$$

where c is free of θ. The most common way to make the likelihood unique, e.g., for the purpose of plotting and comparing, is to set the maximum to one. Once they are normalized to have unit maximum, the width of the likelihood tells us about the amount of information.

What about the third probability axiom on additivity? Suppose $y = 3$ is the outcome of a binomial experiment with $n = 10$ trials and an unknown probability θ. Then we have

$$\begin{aligned} L(\theta = 0.3) &= \Pr(Y = 3|\theta = 0.3) = 0.27 \\ L(\theta = 0.5) &= \Pr(Y = 3|\theta = 0.5) = 0.12. \end{aligned}$$

What is $L(\theta = 0.3 \text{ or } 0.5)$? The definition of likelihood does not tell us what it is or how to compute it. Therefore, the likelihood is not additive. The lack of additivity is a serious weakness that sometimes makes direct likelihood inference fallible. Here's a striking example from Basu (1975).

Ticket paradox. In a box containing 1000 tickets, where 20 are marked with a number θ and 980 with 10θ. You draw a ticket at random and observe the number as $y = 15$. The only non-zero likelihoods of θ are

$$\begin{aligned} L(\theta = 15) &= \Pr(Y = 15|\theta = 15) = 0.02 \\ L(\theta = 1.5) &= \Pr(Y = 15|\theta = 1.5) = 0.98. \end{aligned}$$

Therefore, the MLE is $\widehat{\theta} = 1.5$, which is 49 times more likely than $\theta = 15$. This seems reasonable. But now suppose that the 980 tickets are instead marked with $a_i\theta$, where the constants a_is are 980 known and distinct numbers between 9.9 and 10.1. In effect, these 980 tickets are marked within rounding error of 10θ. Now the non-zero likelihoods are

$$\begin{aligned} L(\theta = 15) &= \Pr(Y = 15|\theta = 15) = 0.02 \\ L(\theta = 15/a_i) &= \Pr(Y = 15|\theta = 15/a_i) = 0.001, \ i = 1, \ldots, 980, \end{aligned}$$

so the MLE is $\widehat{\theta} = 15$, which is 20 times more likely than any θ value around 1.5. However, intuitively these two situations should give a similar inference. There is a cluster of likely values around 1.5, but we cannot combine the information to produce a higher likelihood because the likelihood is not additive.

Although direct use of likelihood for inference is not ruled out, and we show some examples of that, the paradox gives a clear warning: 'Do this at your own risk.' The likelihood value on its own is easily fooled by the parameter space. That's why standard applications of likelihood require probability-based calibration. And, with inductive inference in general, 'external validation' in the form of common sense is needed. Later in Section 12.3 we show how this weakness of the classical likelihood is overcome by the extended likelihood.

The likelihood of a union of parameters needs a new rule. The most commonly used rule is

$$L(\theta = 0.3 \text{ or } 0.5) \equiv \max\{L(\theta = 0.3), L(\theta = 0.5)\},$$

known as the 'profile likelihood.' One of its most useful applications is for removing nuisance parameters. Suppose the probability model is parameterized by a 2 dimensional parameters $(\theta, \eta) \in R^2$, and θ is the parameter of interest. Then, the likelihood of θ alone is conceptually the likelihood of $\cup_\eta(\theta, \eta)$. The profile likelihood of θ is then

$$L(\theta) \equiv L\{\cup_\eta(\theta, \eta)\} \equiv \max_\eta L(\theta, \eta), \tag{10.1}$$

where the maximization over η is performed at fixed θ. This is in contrast to the standard integration to get the marginal probability of y alone from the joint probability density $p(y, z)$:

$$p(y) \equiv p\{\cup_z (y, z)\} = \int_z p(y, z) dz.$$

Why not just add the necessary axioms to turn likelihood into probability? We can do that by

- making it integrate to one in order to satisfy the second probability axiom;
- making it additive so it satisfies the third axiom.

However, one will immediately realize that this process merely reproduces the inverse probability method – i.e., the Bayesian posterior probability – using a uniform prior density. The following paradox indicates that one should be careful before turning or equating likelihood to probability.

The definition of likelihood immediately implies a canonical way of combining information from multiple sources. If the sources are independent, we can just take the product of the likelihoods or, more intuitively, take the sum of the log-likelihoods. For a single dataset y, we can also interpret the likelihood as

$$L(\theta) = L(\theta; y) \times L_0(\theta),$$

where $L_0(\theta)$ is the prior likelihood. In principle, any prior information about θ can be incorporated into $L_0(\theta)$, so $L(\theta)$ represents updated information. If we know nothing about θ, the prior likelihood is *always* a constant free of θ. This follows from the definition, because having no information is equivalent to having irrelevant information whose probability is independent of θ. Thus, the uniform prior likelihood truly represents complete ignorance; unlike the uniform prior probability, there is no paradox associated with the uniform prior likelihood.

10.2 Exchange Paradox

The recognition that likelihood is not probability is crucial for solving a tricky paradox that has puzzled many extremely bright people, including those we know personally. It's the so-called exchange paradox, which goes like this:

> An unknown θ dollars is put in one envelope and 2θ dollars in another. You are asked to pick one envelope *at random*, open it, and then decide if you would exchange it with the other envelope. So, you pick one and open it to reveal $y = \$100$. Now you reason: suppose M is the amount in the other envelope, then M is either \$200 or \$50 with equal chance. If you exchange your envelope, the expected value of your winning is $(200 + 50)/2 = 125 > 100$. So, you would exchange the envelope, wouldn't you? Since this is true for any value of y, *there is no need to open the envelope, and you would still want to exchange the envelope*, which is of course absurd, since you have just chosen it randomly!

Also known as the two-envelope paradox, a quick read in Wikipedia shows that the exchange paradox has an extensive literature, offering numerous explanations. The source of the paradox is the reasoning based on probability, which leads to averaging. If we *knew absolutely nothing* about θ, then on observing $Y = 100$ we can only have $\theta = 100$ or $\theta = 50$ with likelihoods

$$\begin{aligned} L(\theta = 50) &= \Pr(Y = 100|\theta = 50) = 0.5 \\ L(\theta = 100) &= \Pr(Y = 100|\theta = 100) = 0.5. \end{aligned}$$

That is, the allowed θ values have the same likelihood, not the same probability. Consequently, the other envelope is either 50 or 200 with equal likelihood, not probability. But we do not have a rule on how to use likelihood as a weight for averaging. Therefore, there is no rational way to prefer one envelope over the other, and the paradox is avoided. Yet, here the likelihood satisfies the psychological need to attach some uncertainty to a unique case, something denied by the frequentists. However, there are many more twists in this wonderful paradox; we shall analyse it in more detail in Chapter 19.

10.3 Prosecutor's Fallacy

The so-called prosecutor's fallacy also arises out of the confusion between probability and likelihood. The application of statistical inference in court has been the subject of serious discussion and debate, especially after the emergence of DNA profiling as part of forensic evidence. The logic of legal concepts such as 'presumed innocence' or 'guilt beyond reasonable doubt' has direct statistical connotations, so the principles apply more generally to any assessment of evidence. As discussed in Gardner-Medwin (2005), three key propositions are at issue:

A: The facts or evidence could have arisen if the defendant is guilty.
B: The facts or evidence could have arisen if the defendant is innocent.
C: The defendant is guilty.

Clearly, A and not-B together would imply C, but C does not imply not-B. The latter is obvious if the evidence is weak, i.e., could easily have been found among innocent people. Thus, strong beliefs in A and not-B together are a more stringent requirement than a belief in C alone. In fact, for expert witnesses, the presumed innocence requirement may preclude the assessment of C. Since the categorical truth of these statements is in reality rarely available, the prosecution may have to present extremely small probabilities to establish guilt beyond reasonable doubt.

To be more specific, suppose there is a successful match between a defendant's DNA and the crime-scene sample DNA. We have two natural parameters: θ_0 is the hypothesis that the defendant is innocent, and θ_1 that the defendant is guilty. (More correctly, θ_0 means the defendant is not the source of the crime sample, because 'innocent' or 'guilty' is for the jury to decide. But we use those words here for their more suggestive effects. They do not affect the statistical reasoning.) On the basis of the data, we can compute

$$L(\theta_0) = \text{Pr}(\text{DNA match}|\text{innocent}).$$

This likelihood is equal to the probability that a random person will have DNA that matches the DNA from the crime scene sample. This assesses the evidence for Proposition B above. With additional assumptions and statistical information about the distribution of the DNA markers in the background population, one can compute this probability. In forensic applications, DNA markers are chosen such that this likelihood is very small, say 10^{-9}, thus supporting not-B, while the comparative likelihood is assumed to be certain:

$$L(\theta_1) = \Pr(\text{DNA match}|\text{guilty}) = 1.$$

This means that the evidence of a DNA match is highly specific evidence and points to the defendant as the source, i.e., strongly supporting A. Therefore, these likelihoods provide an assessment of Propositions A and B above. The level of evidence is often reported in terms of the likelihood ratio

$$R = L(\theta_1)/L(\theta_0),$$

typically a very large number, such as 10^9, as evidence against the defendant.

So, likelihood-based reasoning leads to a likelihood ratio R; this number provides the joint assessment of the propositions (A and not B). In contrast, the conditional probabilities are

$$\begin{aligned} P_0 &= \Pr(\text{innocent}|\text{DNA match}) \\ P_1 &= \Pr(\text{guilty}|\text{DNA match}). \end{aligned}$$

The so-called prosecutor's fallacy is to misrepresent the likelihood $L(\theta_0)$ as $\Pr(\theta_0|\text{DNA match})$. The probability ratio is

$$\frac{P_1}{P_0} = \frac{L(\theta_1)}{L(\theta_0)} \times \frac{p_1}{p_0} = R \times R_0,$$

where $R_0 \equiv p_1/p_0$ is the ratio of prior probabilities of guilt vs innocence.

One may argue that probability-based reasoning based on P_1/P_0 is better, but in fact the same result can also be achieved within likelihood-based reasoning. It just requires a different interpretation of R_0 as the prior likelihood ratio of guilt vs innocence, so the total likelihood ratio is the same $R \times R_0$.

The practical issue in court is the establishment of prior likelihoods/probabilities of guilt that can be agreed upon by all parties. One can easily imagine the contentious arguments on how to set the priors. For example, is it reasonable to presume that the defendant is a random person from the general population? If so, then the prior p_1 should be small. However, the prosecutor could argue that the defendant is a known criminal, so p_1 should be much greater than p_1 of a random person. Then, how do we abide by the presumed innocence requirement?

In summary, the prosecutor's argument does not have to rely on a probability-based argument. A valid likelihood-based reasoning depends on the likelihoods alone, hence avoiding the fallacy. Unfortunately, in layman's language, probability and likelihood are interchangeable expressions of uncertainty, thus confusing the two valid modes of reasoning and making it difficult to avoid the prosecutor's fallacy. We shall continue the discussion of forensic applications in Section 17.3.

10.4 Likelihood Principle

Why should one use the likelihood in statistical inference? The definition of likelihood does not tell us its fundamental role in statistical inference. One can appeal to the desirable large-sample theoretical results: (i) In large samples the MLE converges to the true value (this is called the consistency property), and (ii) it achieves the best precision among consistent estimates of θ. But there are also qualitative results, almost axiomatic in their power, that tell us that the likelihood contains all the evidence about θ in the data. So, the use of anything other than likelihood might incur a loss of information.

In 1922, Fisher introduced the notion of a sufficient estimate to mean that it 'summarize[s] the whole of the relevant information supplied by the sample.' He conjectured optimistically that the method of maximum likelihood would automatically lead to sufficient estimates. This was later shown to be untrue, for instance, in location families such as the Cauchy model (Fisher 1934). In current statistical theory, the concept is usually attached to a general statistic rather than to an estimate. A sufficient estimate may not exist, but a sufficient statistic always does.

As a statistic, i.e., as a function of the data, the likelihood function is minimal sufficient, in the sense that it is a function of any other sufficient statistic (Pawitan, 2001, Section 3.2). To get a sense of the minimality property, note that computing a function of the data potentially reduces the amount of information. For example, by reporting or keeping only $y_1 + y_2$, we reduce the amount of data from a pair of values (y_1, y_2) to just its sum, so we lose some details. Thus, minimal sufficiency means maximum reduction without loss of information. The MLE may not capture all the information in the data, but the likelihood function does, so we still have to start with the likelihood.

Birnbaum's theorem

A different kind of qualitative support for likelihood comes from the controversial and poorly understood likelihood principle. The principle is based on a fundamental theorem proved by Alan Birnbaum in 1962, the year Fisher died. Let E be an experiment defined by a triplet $\{Y, \theta, p_\theta(y)\}$. An abstract evidence $\text{Ev}(E, y)$ about θ is defined as a function of the experiment and the observed data $Y = y$. Without thinking of anything specific, what properties should $\text{Ev}(\cdot)$ have? Birnbaum defined two axiomatic requirements that $\text{Ev}(\cdot)$ must satisfy and called them the sufficiency and conditionality principles. We have already met the former. A statistic $t \equiv t(y)$ is called sufficient if the conditional probability $p_\theta(y|t)$ is free of θ. This means that, as far as θ is concerned, once we know $t(y)$, there is no more information left in the data. So, first, we have the following:

> **Sufficiency principle (SP):** If y_1 and y_2 are two datasets from an experiment E such that $t(y_1) = t(y_2)$, then $\text{Ev}(E, y_1) = \text{Ev}(E, y_2)$.

This is highly sensible: If the sufficient statistic captures all the information about θ, then different datasets with the same sufficient statistic should carry the same evidence about θ. Statisticians of all stripes accept this principle.

The second requirement is a bit more subtle, relying on a mixture experiment. Suppose there are two experiments $E_1 = \{Y_1, \theta, p_{1,\theta}(y_1)\}$ and $E_2 = \{Y_2, \theta, p_{2,\theta}(y_2)\}$, where only the parameter θ needs to be common. Now consider a mixture experiment E^*, where first a coin is tossed, so that a random number $J = 1$ or 2 is generated with probability 0.5 each; the value of the probability does not matter as long as it is free of θ. The experiment E_j is then performed conditionally on the observed j. Formally, the outcome from the mixture is $Y^* = (J, Y_J)$ with realized outcome $y^* = (j, y_j)$. Then we have the following:

Conditionality principle (CP): The evidence from a mixture experiment is equal to the evidence from the experiment actually performed, i.e., $\text{Ev}(E^*, y^*) = \text{Ev}(E_j, y_j)$.

When one considers most experiments performed in real life, they are typically affected by uncontrollable external factors that make the actual experiment look like a realization of a mixture experiment. For example, the number of participants in a clinical trial will vary according to recruitment effort or funding, etc.; lab experiments may be affected by random batch effects, personnel, materials, equipment, environmental factors, etc. The CP implies that the evidence should be conditional upon the performed experiment, not on experiments that could have been performed. This matters because an unconditional analysis based on the mixture distribution of Y_1 and Y_2 could produce different results from the conditional analysis (Pawitan, 2001, Section 7.3).

The general version of the coin toss to decide the experiment is called an ancillary statistic, which is a statistic whose distribution is free of the parameter θ but its value is informative about precision. The sample size of a study is a common example of an ancillary statistic, but there are many others; we shall return to this topic in Chapter 13. In frequentist conditional inference (e.g., Reid 1995), conditioning on the ancillary statistic is typically done to make the inference *more relevant to the data at hand.* There is a hint of evidential purpose here. So, at least in the first reading, the CP also sounds highly sensible. But, wait, now consider:

Strong likelihood principle (SLP): Suppose y_1 and y_2 are datasets from experiments E_1 and E_2, respectively, such that the likelihood functions are equivalent, i.e.,

$$L(\theta; y_1) \sim L(\theta; y_2).$$

These two datasets then carry the same evidence: $\text{Ev}(E_1, y_2) = \text{Ev}(E_2, y_2)$.

If y_1 and y_2 are from *the same experiment*, then the equality of sufficient statistics $t(y_1) = t(y_2)$ implies equivalent likelihoods. By SP, the datasets also have the same evidence; therefore, two datasets with equivalent likelihoods have the same evidence. Thus, the SP is also called the weak likelihood principle (WLP). So, to be non-trivial, when the term 'likelihood principle' without qualifier refers to the SLP, which allows the data to come from *different experiments.*

To highlight the difference between the WLP and SLP, let's consider this classic example: Suppose we have (i) a (tiny) dataset comprising y_1 successes from a binomial experiment with n trials is fixed in advance, and (ii) another dataset $y_2 = n$ trials observed in a negative-binomial experiment with the number of successes y_1 is fixed in advance. Then

$$\begin{aligned} L(\theta; y_1) &= \begin{pmatrix} n \\ y_1 \end{pmatrix} \theta^{y_1}(1-\theta)^{n-y_1} \\ L(\theta; y_2) &= \begin{pmatrix} n-1 \\ y_1-1 \end{pmatrix} \theta^{y_1}(1-\theta)^{n-y_1} \\ &\sim L(\theta; y_1). \end{aligned}$$

Therefore, these two experiments/datasets produce equivalent likelihoods and, according to the SLP, they carry the same evidence about θ. Many statisticians find this unacceptable, or at least not self-evident. But that's not really an issue yet, because you could just reject the SLP outright. The surprising theorem from Birnbaum states:

Theorem: The SP and CP together are equivalent to the SLP.

Thus, if you accept the more self-evident SP and CP, then you must also accept the SLP. A key implication of the SLP is that $\text{Ev}(E, y)$ depends on E and y only through the likelihood function or that the likelihood contains all the evidence in the data. Although the SP and CP axioms are defined for abstract evidence, the theorem implies that the evidence that satisfies SP and CP is *embodied in the likelihood function.* As a mathematical result, this should not be controversial.

10.5 Controversy

Alas, the likelihood principle is controversial, particularly among frequentists. What generates the controversy? It starts with a normative interpretation of the SLP (see, e.g., Cox and Hinkley, 1974, Section 2.3: Bjørnstad, 1996; Mayo, 2014) that says

> two datasets – regardless of the experimental source – with equivalent likelihoods *should lead to the same inferences or conclusions.*

The problem is that two datasets with equivalent likelihoods – e.g., the binomial and negative binomial data above – can have distinct P-values or confidence intervals. Therefore, standard frequentist statistical inferences, such as P-value or confidence procedures, are in violation of *this version* of the SLP. In order to contrast it to the original empirical-evidential SLP, let's denote this normative version of the SLP by iSLP, as it refers to inferences. In the iSLP, the abstract evidence $\mathrm{Ev}(E, y)$ is replaced by a distinct but still abstract concept of 'inference' $\mathrm{Inf}(E, y)$.

One clearly unacceptable application of the iSLP is in sequential experiments with optional stopping. The negative-binomial experiment is one such experiment where, at each step, one simply checks whether the current number of successes has reached a pre-set goal. In the general case, the data are collected sequentially so that at each step, data collection may stop with a probability that depends only on the observed data and not on the unknown parameter or any unobserved data. Then we can show that the stopping rule does not affect the likelihood; a proof is given in this chapter's appendix. According to the iSLP, statistical inferences should then ignore the stopping rule.

However, optional stopping clearly affects the long-term statistical performance of, for instance, test procedures. Stopping while you are ahead, which does not affect the likelihood, can easily and substantially increase the significance level of a test. Thus, no frequentist statistician will follow the iSLP. Unfortunately, instead of simply questioning the normative interpretation of iSLP, the controversy engulfs the whole SLP concept. Here is an example from Armitage (1961); a full description and a numerical example are given in Pawitan (2001, Chapter 7).

Stopping rule paradox. Suppose we observe $y_1, y_2, \ldots$ sequentially and independently from $N(\theta, 1)$, where we're interested to test the null hypothesis H_0: $\theta = 0$ vs H_1: $\theta \neq 0$. The stopping rule is to check at each step and stop when

$$z_n \equiv \sqrt{n}\bar{y}_n \geq 1.96,$$

where $\bar{y}_n$ is the average at step n. Because z_n is the standard z-test statistic, in effect, we collect data until H_0 is rejected at the 5% level. Theoretically, under H_0, this sequential procedure is guaranteed to stop, at which point the test is significant. So we're gathering data to a foregone conclusion; the true type I error probability of the procedure is 100%. Yet, the likelihood is not affected by the optional stopping; it's the same as if the sample size had been decided in advance.

The problem with the iSLP is that inference – as in 'making an inference' – is a proactive concept, which makes the iSLP normative, telling you what to do. In contrast, evidence is a passive entity, so the SLP is only descriptive; it does not tell you what to do. Indeed, virtually no statistician draws an inference directly from the likelihood, so equivalent likelihoods will not make them feel obliged to make the same inferences. Bayesians will supplement the likelihood with a prior, while frequentists will consider the sampling properties or the data generating process. Both the prior and the sampling properties can be context-specific and can differ in different experiments.

Yet, and this is perhaps the confusing part in the controversy, Birnbaum's theorem *holds* for the iSLP, because, in the formal proof, we can just replace the abstract notation $\text{Ev}(\cdot)$ with $\text{Inf}(\cdot)$. This means that the iSLP is equivalent to the iSP and iCP together, the corresponding inferential versions of the SP or CP, but they have potentially different interpretations. So, if we reject the iSLP then we must also consider the iSP and iCP.

The SP refers to equal sufficient statistics *from one experiment*, so it is self-evident that they should lead to the same inferences. The iSP is thus logically equivalent to the original SP. Therefore, if we find the iSLP unacceptable, it must be due to the iCP, which is indeed logically very different from the CP. How we perform statistical inference in a mixture experiment depends on the context of the experiment and the goal of the inference. As we discuss next, if we care about the long-term properties, an unconditional analysis could be preferable to the conditional one, thus violating the iCP.

Analysis of 2-by-2 tables

Statistical inference for 2-by-2 tables provides an important illustration, as it is one of the most commonly performed analyses in statistics. Yet, there is a long controversy whether one should use a conditional or an unconditional analysis. We guess that the issue is confusing enough for statisticians and worse for non-statisticians.

Thus, suppose that we want to compare two binomial proportions: the proportion of sick subjects in an exposed group and a control group. The numbers of subjects in each group (n_1 and n_2) are naturally fixed by the study plan/design, but the corresponding numbers of sick subjects

(y_1 and y_2) are random. Let's denote the total number of sick subjects by $n_y = y_1 + y_2$. An exact conditional analysis is based on fixing n_y at the observed value. The effect is to have both margins of the 2-by-2 table fixed; the resulting test is known as Fisher's exact test. In contrast, an exact unconditional analysis does not make the conditioning step; the resulting test is given, for example, by Barnard (1947) or Suissa and Schuster (1985). Which is better?

Let's start with an easier issue. Some authors, especially from the non-statistical side (e.g., Ludbrook, 2013), argue that Fisher's exact test 'requires the rare condition that both row and column marginal totals are fixed in advance,' so it's not natural as the data actually come with only one margin fixed. (Fisher's exact test also applies to multinomial samples, where both margins are free.) This sounds like a legitimate criticism, but it is actually a theoretical misunderstanding. A conditional test *does not* imply that the inference applies only to data with the same margins as the observed margins; indeed, the inference applies to the original unconditional data. This can be seen in the extreme case of the permutation test (Section 11.1), where we fix the whole observed data, but it's only a theoretical device to derive the null distribution. The conditioning step does not limit the generality of the test to data that resemble the observed data.

All proponents of the unconditional test argue legitimately that it has a higher statistical power than the conditional test. The loss of power in the conditional test occurs because conditioning reduces the number of points in the sample space. But, in inference, power is not the only property that matters. In 1955, Fisher wrote how Barnard initially presented his unconditional test as 'much more powerful than Fisher's,' but

> after some discussion, [Barnard] had the generosity to go out of his way to explain that further meditation had led him to the conclusion that Fisher was right after all.
>
> Professor Barnard has a keen and highly trained mathematical mind, and the fact that he was misled into much wasted effort and disappointment should be a warning that the theory of testing hypotheses set out by Neyman and Pearson has missed at least some of the essentials of the problem, and will mislead others who accept it uncritically.

Why did Barnard (1949) change his mind? It's due to the conditionality principle. If we tell you that the total number of sick subjects $n_y = 127$, can you tell us what the difference is between the proportions of sick people in each group? Intuitively, you can't. This means that n_y is ancillary; it behaves as the coin toss in the mixture experiment, so a proper analysis should be conditional on n_y. That's Fisher's argument that convinced Barnard to change his mind. (Theoretically, n_y is only an approximate ancillary. If the sample sizes n_1 and n_2 differ substantially, there is a slight information in n_y on the difference between the groups. See Choi et al, (2015) for further discussion.)

So, if we care about the evidence in a particular experimental result as presented in a 2-by-2 table, the conditional analysis is more appropriate than the unconditional one. On the other hand, if we happen to deal with an application where we regularly collect 2-by-2 tables and we care about long-term performance but not about individual tables, then the unconditional analysis is more appropriate, thus violating the iCP. To emphasize, even in this latter situation, *the evidence* in any particular table is still captured by conditional likelihood, which means that the CP holds. But, unlike the iCP, the CP does not oblige us to make inferences based solely on that likelihood.

Appendix: Sequential Experiment with Optional Stopping

Here we show that optional stopping does not affect the likelihood. Suppose that y_i is collected sequentially and independently for $i = 1, 2, \ldots$, where at each time t we make a probabilistic stopping rule whether or not to continue, depending only on the data collected so far. Let's call this probability $h_t(y_1, \ldots, y_t)$. When the experiment is stopped, call the sample size n, so at that point we have the data $y \equiv (y_1, \ldots, y_n)$. We can always guarantee that the experiment will stop by setting $h_T(\cdot) \equiv 0$ at some desired maximum time point T.

For the negative binomial experiment, the stopping rule is simple: Continue (i.e., $h_t(\cdot) = 1$) if the running total $\sum y_i < k$ for a pre-specified k, and stop when $\sum y_i = k$.

Given a model $p_\theta(y_i)$, the likelihood function is

$$\begin{aligned} L(\theta; y) &= \text{Pr(deciding to observe } y_1)p_\theta(y_1) \times \\ &\quad \text{Pr(deciding to observe } y_2|y_1)p_\theta(y_2) \times \cdots \times \\ &\quad \text{Pr(deciding to observe } y_n|y_1, \ldots, y_{n-1})p_\theta(y_n) \times \\ &\quad \text{Pr(stop at } y_n|y) \\ &= h_0 p_\theta(y_1) \times \\ &\quad h_1(y_1)p_\theta(y_2) \times \cdots \times \\ &\quad h_{n-1}(y_1, \ldots, y_{n-1})p_\theta(y_n) \times \\ &\quad (1 - h_n(y)) \\ &= \text{constant} \times \prod_i p_\theta(y_i), \end{aligned}$$

which means that the stopping rule does not affect the likelihood. In other words, the likelihood is the same as if we had collected the data without any sequential arrangement at all.

11

P-values and Confidence

11.1 Significance Test via Randomization

Among the many groundbreaking concepts that Fisher developed in the 1920s, significance testing and the P-value are perhaps the most influential in applied research. Let's consider a simple example. Suppose that we have a dataset on the nicotine content of the two brands A and B of pipe tobacco in Table 11.1.

We want to test the null hypothesis H_0: The nicotine content of the two brands is the same. The t-test provides the P-value 0.0163. For this small sample, the t-test is valid only when the population follows the normal distribution. But the data have many ties, so the normal assumption is doubtful. Let the test statistic be $Y = \bar{D} = \bar{X}_1 - \bar{X}_2$. Here, the total sample size $n = n_1 + n_2 = 19$ is too small to use the normal approximation for $\bar{D}$. What is our avenue for a trustworthy analysis?

A significance test based on randomization or permutation was proposed by Fisher (1935), who used the classic tea-tasting experiment to motivate it. Now, associated with each brand, think of a population of values with some unspecified distribution, not necessarily normal. If H_0 is true in the sense that the two populations are the same (not just in the mean), the data are random samples from a single population. Intuitively, under H_0, the tobacco brands A and B behave as if they were just randomly permuted labels among the 19 sample values, with the observed data being a particular permutation.

Alternatively, we can randomly re-assign the 10 observations *without replacement* as A and the rest as B. Here we have $\binom{19}{10} = 92,378$ possible values of test statistics from those possible samples. The data observed in the table is one particular random sample among the 92,378 randomization samples. Figure 11.1(a) shows the histogram of the distribution of $\bar{D}$ of the 92,378 randomization samples. From the data, we have the

TABLE 11.1
Nicotine content (mg/g) in 19 samples from two brands of pipe tobacco.

A	24	24	26	25	27	29	27	25	27	23
B	22	18	26	16	25	28	19	24	14	

observed value of the test statistics $y = |\bar{d}| = 25.70 - 21.33 = 4.37$, which gives the P-value, defined as

$$\text{P-value} \equiv \Pr(Y \geq y|H) = \Pr(|\bar{D}| \geq |\bar{d}||H) = \frac{1605}{92378} = 0.0174.$$

Thus, Fisher suggested that the null hypothesis can be rejected at the significance level $\alpha = 0.05 > 0.0174$.

If we take 10 random samples *with replacement* for tobacco A, and 9 random samples with replacement for tobacco B from the 19 observations, they will be called bootstrap samples. The bootstrap samples can be arbitrarily large. Figure 11.1(b) shows the bootstrap distribution based on the 10^5 bootstrap samples to give the bootstrapped P-value

$$\Pr(|\bar{D}_{boot}| \geq |\bar{d}||H) = \frac{1743}{10^5} = 0.0174,$$

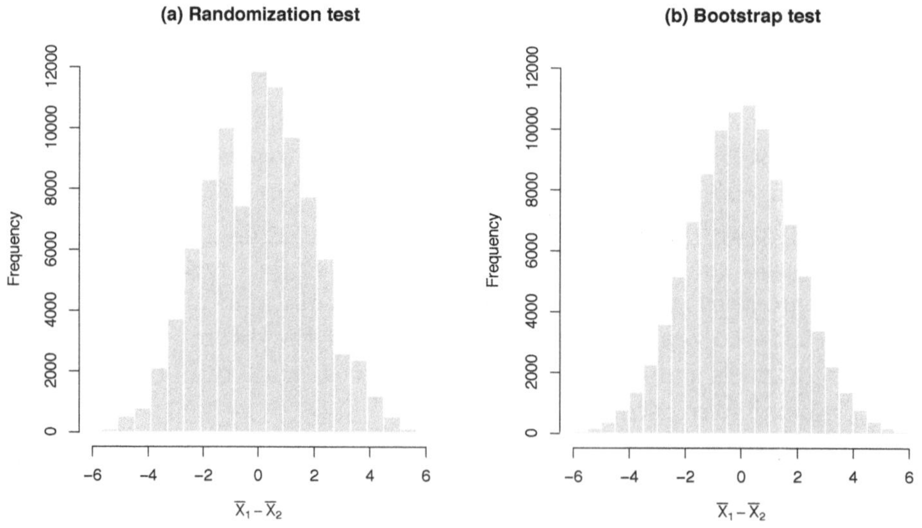

FIGURE 11.1
(a) Histogram of $\bar{X}_1 - \bar{X}_2$ from the randomization test of the nicotine data in Table 11.1. (b) The same for the boostrap test.

which is equal to that of the randomization test to 4 decimal places. Thus, randomization and bootstrapping are useful for computing the P-value in general. The similarity between the randomized and bootstrap tests can be expected even in just moderate-sized samples such as in this example. The advantage of the randomized test is that the P-value is correct even in very small sample sizes, whereas the bootstrap requires larger samples. However, the bootstrap idea may apply more generally where the randomized test may not work.

11.2 P-values and Fiducial Probability

Suppose someone claims that he has a special way of tossing a coin to make the chances of heads greater than 1/2. To demonstrate this, he tosses the coin 5 times and gets 5 heads. Is that convincing evidence? If that sounds convincing, is it more convincing or less convincing than getting 9 heads out of 10 tosses? You might guess the latter is stronger evidence, but how would you establish that quantitatively? The answer is with the P-value. What we want to test is the hypothesis $H : p = 1/2$, based on observing $y = 5$ heads in $n = 5$ tosses, or $y = 9$ in $n = 10$ tosses. For the former, the (one-sided) P-value is defined as

$$\begin{aligned}\text{P-value} &\equiv \Pr(Y \geq y|H) \\ &= \Pr(Y \geq 5|p = 1/2) \\ &= 1/32 = 0.0312,\end{aligned}$$

when Y is binomial with parameter $n = 5$ and $p = 1/2$. And, for the latter

$$\text{P-value} = \Pr(Y \geq 9|H) = 0.0107.$$

A smaller P-value indicates stronger evidence. So, the latter is indeed more convincing evidence for the claim.

The P-value is still one of the most commonly used statistical tools in scientific research, though there are dissenting views about its utility. However, other more generally acceptable inferences can start with the P-value as a function of the parameter. That is, instead of just thinking of a specific hypothesis $p = 1/2$, let's consider the function

$$F(p) \equiv \Pr(Y \geq y = 5|p) = p^5,$$

for all p between 0 and 1. Clearly, $F(p)$ is strictly increasing in p, $F(0) = 0$, and $F(1) = 1$, so it behaves exactly like a probability distribution function. Fisher (1930) called this the *fiducial distribution* of p. It is a term loaded with a history of controversies, but first consider its utility:

- It can be simply seen as a collection of P-values; for instance, $F(1/2)$ is the P-value for testing the hypothesis H; $p = 1/2$, but in principle we can test any p.

- *More importantly*, we can easily obtain confidence intervals using $F(p)$ as a lookup table. For instance, to get a 95% confidence interval for p, we can see that $F(1)-F(0.55) = 0.95$, so the CI is $0.55 \leq p \leq 1$.

Fisher's (1930, 1933) main motivation was that we can get all these inferences *without* having to specify any prior probability on p as required in the classical inverse probability method. Technically, he introduced the fiducial concept based on a continuous and sufficient estimate, so it is possible to get exact results. But in its current usage as 'confidence distribution' there is no such restriction; see more below.

The concept of fiducial probability ran into immediate problems because Fisher claimed that it could be treated as ordinary probability. In his highly polemical paper, Basu (1975) 'vividly' recalled a 1955 seminar by Fisher at the Indian Statistical Institute in Calcutta, where he gave a syllogistic form of the fiducial argument. Suppose X is known to be normally distributed with unit variance and unknown mean θ; the only thing known about θ is that $-\infty < \theta < \infty$. Then he claimed:

> *Major premise:* Probability that the variable X exceeds θ is 1/2.
> *Minor premise:* The variable X is observed and the observation is 5.
> *Conclusion:* Probability that θ is less than 5 is 1/2.

Needless to say, this is very confusing. It might have helped if he had used Carnap's notation of Probability$_2$ (= frequentist probability) for the major premise and Probability$_1$ (= epistemic probability) for the conclusion, but then again, he wanted the two probabilities to be the same 'ordinary probability.' It's somewhat mysterious why Fisher never recognized or accepted that the fiducial probability is not a standard (Kolmogorov) probability.

What makes the fiducial concept so tempting is that for standard inferences of its 'native' parameter space – the parameter used in its construction via the P-value – the fiducial distribution enticingly behaves like a probability distribution. Moreover, from its construction, we can also see that one-to-one transformations of the parameter follow the standard rule of probability. However, while fiducial probability trivially satisfies the first and second probability axioms, it does not fully satisfy the crucial third axiom on additivity. The lack of additivity becomes conspicuous when dealing with many-to-one transformations.

Counterexamples are easy to construct: Suppose an observation $y = 0.5$ is taken from $N(\mu, 1)$. Then the fiducial distribution of μ is

$$F(\mu) = \Pr(Y \geq 0.5|\mu) = 1 - \Phi(0.5 - \mu) = \Phi(\mu - 0.5),$$

where $\Phi(\cdot)$ is the standard normal distribution function. That is, $F(\mu)$ is the normal distribution centred at 0.5. This is actually the best example of a fiducial distribution: All standard inferences about μ – such as P-values and confidence intervals – can be conveniently and intuitively gleaned from this fiducial distribution.

Now, what is the fiducial probability of $\theta \equiv \mu^2 < 1$? Treating $F(\mu)$ as an ordinary probability distribution, the fiducial probability of $\theta < 1$ should be equal to

$$F(1) - F(-1) = \Phi(0.5) - \Phi(-1.5) = 0.62.$$

And for the small region $\theta < \epsilon$ near zero, the fiducial probability of $\theta < \epsilon$ goes to zero as $\epsilon \to 0$, so there is no point mass at zero.

But, following its definition, the correct fiducial probability should be computed based on Y^2, because Y^2 is χ_1^2 with the non-centrality parameter θ (so technically Y^2 is sufficient for θ). Thus, on observing $y = 0.5$, the fiducial distribution of θ is

$$\begin{aligned} G(\theta) &= \Pr(Y^2 \geq 0.25|\theta) \\ &= 1 - \Phi(0.5 - \sqrt{\theta}) + \Phi(-0.5 - \sqrt{\theta}). \end{aligned}$$

In particular, the fiducial probability of $\theta < 1$ is equal to

$$G(1) = \Phi(0.5) + \Phi(-1.5) = 0.76$$

and, as ϵ goes to zero, $G(\epsilon)$ goes to $1 - \Phi(0.5) + \Phi(-0.5) = 0.62$. This means $G(\theta)$ has a point mass at zero. Thus, completely different from the results based on $F(\mu)$. The problem arises because, although the interval $[\theta < 1]$ is equivalent to $[-1 < \mu < 1]$, the steps required to obtain the fiducial probabilities are not compatible.

In general, a lack of additivity generates marginality problems, which are a serious issue as finding a marginal distribution is inevitable in multiparameter cases. For instance, we can construct a joint fiducial distribution $F(\mu, \sigma)$ of the normal mean μ and the standard deviation σ. But there is no guarantee that we can obtain a correct fiducial distribution for new parameters such as μ/σ using the standard probability calculus from $F(\mu, \sigma)$. A well-known failure of fiducial theory is its failure to provide an exact solution to the so-called Behrens-Fisher problem, which is how to construct inference for the difference of two normal means $\mu_1 - \mu_2$, when the two variances σ_1^2 and σ_2^2 are both unknown and unequal. We note, however, that similar marginalization paradoxes are also an issue with objective Bayesian posteriors; see Section 9.4.

11.3 Confidence Procedures and Coverage Probability

Fisher originally thought that his fiducial probability was an objective probability satisfying ordinary probability laws, but that's clearly not the case. In later papers, he wrote that it represents the epistemic state of mind, consistent with the probability of the 'early writers.' In the classification of probability theories in Chapter 6, we might consider his main goal as to establish Keynesian logical probability, which is a degree of rational belief among intelligent individuals. The fiducial probability would also replace the Bayesian subjective probability, which presumed an arbitrary prior probability. Although not for lack of trying, Fisher never managed to convince others of the merits of the fiducial argument. It is considered his biggest blunder and has been largely ignored in practical statistics today.

However, the fiducial distribution and the intervals derived from it motivated Neyman's *confidence procedure* (Neyman, 1937). After all, the word *fiduciam* in Latin means 'confidence.' The confidence procedure became the one universally accepted, perhaps because it is much less ambitious and carries much less philosophical baggage than the fiducial idea. Instead of trying to solve the inductive inference problem, Neyman simply focused on the deductive properties of the confidence intervals. A 95% confidence interval $L_1(Y) < \theta < L_2(Y)$ only needs to satisfy one primary requirement: In the long run, it is correct 95% of the time. That is,

$$\Pr_\theta(L_1(Y) < \theta < L_2(Y)) = 0.95.$$

The 95% probability is called the confidence level, and the probability is called the coverage probability. Once y is observed, we'd say we have 95% *confidence* in the interval $L_1(y) < \theta < L_2(y)$. However, unlike Fisher, Neyman was an avowed frequentist. He was strongly against attaching the confidence level as a degree of uncertainty to the *observed interval.* So Neyman's 'confidence' actually has the same meaning as probability. There are subsequent theoretical results that establish the optimality properties of commonly used confidence procedures, such as the t-interval. But, philosophically, the orthodox frequentist interpretation from Neyman still holds today.

How should we interpret the confidence level or the coverage probability? Orthodox teaching in most statistics textbooks provides an objective frequentist interpretation, which is a mixture of von Mises's frequency theory and Popper's propensity (Chapter 6). For example, a 95% confidence interval means that if we repeat the procedure 100 times, then 'approximately' 95% of the 100 produced intervals will include the true unknown value. And, the 95% is said to apply to the procedure, not to any observed interval.

Suppose that a 95% confidence interval of the cure rate of a certain disease is reported as 50% to 60%. The frequency interpretation is that if we repeatedly make 100 confidence intervals, based on 100 independently generated datasets from the same model of $\Pr(\text{Data}|\theta)$, then among the 100 intervals, around 95% will cover the true cure rate and 5% will not. This is scientific operationalism applied to the confidence interval procedure. If we know the data generation scheme, we can repeat the procedure indefinitely in thought experiments, so that we can obtain the frequency of the coverage rate of the procedure. The frequentists face a quandary: On the one hand, no scientist would repeat the same experiment over and over, so this coverage probability applies only to the outcomes of a thought experiment. On the other hand, the scientist needs an objective measure of uncertainty for the experiment actually performed, and the confidence interval is the only accepted one.

Instead of the long-term frequency interpretation, what if we are interested in the epistemic interpretation of the cure rate between 50% and 60%? Instead of saying that we will be correct in 95 out of

100 experiments, we would like to claim that this interval is correct with 95% confidence, which describes the state of mind.

To write it more explicitly, suppose that the confidence interval $CI = [L_1, L_2]$ is reported. Let

$$U \equiv I(L_1 \leq \theta \leq L_2) = I(\theta \in CI),$$

where $U = 1$ if the interval covers the true θ, and 0 otherwise. Thus, U (as a function of data) is a binary random variable with coverage probability

$$\Pr(U = 1) = 95\%.$$

This coverage probability is an objective probability of the confidence interval if we repeat the same experiment over and over again. But, given *the observed interval*, we want an epistemic confidence

$$C(u = 1) = 95\%,$$

rather than coverage probability of the interval, which requires a repetition of the same experiments. Let E be the proposition that the specific interval $[\ell_1, \ell_2]$ covers the true parameter θ and $u = I(E)$. But, without any other explanation, this 'confidence' of course sounds only like a linguistic trick. A full explanation of the concept of epistemic confidence is discussed in Chapter 13.

Confidence distributions and densities

Although Neyman's confidence procedure is considered the theoretical basis for the confidence interval that we use today, the interval itself can always be calculated from the fiducial distribution. The fiducial distribution is a statistic that contains rich information for frequentist inference. So in recent years (e.g., Schweder and Hjort, 2016), the fiducial distribution has actually been resurrected as a respectable concept 'confidence distribution.' Because the concept of confidence is not controversial, neither is the confidence distribution, as long as it is not considered a probability distribution. However, as with the fiducial distribution previously, the confidence distribution has additivity properties in its 'native' parameter space used in its construction via the P-value.

Given the confidence distribution $F(\theta)$, the confidence density is defined by

$$c(\theta) = dF(\theta)/d\theta.$$

We can generate confidence intervals most intuitively using the confidence density. Suppose that y is an observation taken from a normal distribution with mean θ and variance 1. The confidence distribution is

$$F(\theta) = \Phi(\theta - y),$$

where $\Phi(\cdot)$ is the normal distribution function centred at y. So, the confidence density is

$$c(\theta) = \phi(\theta - y),$$

the normal density centred at $\theta = y$. As before, let's define U as the coverage status of a CI derived from the confidence density. For normally distributed data, the confidence $C(u = 1)$ of the observed interval is the same as the coverage probability $\Pr(U = 1)$. Thus, the degree of belief in the confidence interval in the world of the mind is the same as the coverage probability in the real world. The normal model is of special importance because it generally holds as a large-sample approximation.

11.4 Can We Say with Complete Confidence that the Sun Will Rise Forever?

Going back to the sunrise problem (Section 8.5), let E be the proposition 'The sun will rise tomorrow' and G 'The sun rises forever.' Previously, using the Bayes-Laplace formula with uniform prior, we get $\Pr(G|\text{Data}) = 0$ no matter how much data we have in terms of n consecutive days of sunrises. We can call that a probability dilution effect due to the consideration of an infinite number of future instances. Recall Bayes's formula

$$\Pr(G|\text{Data}) = \frac{\Pr(\text{Data} \cap G)}{\Pr(\text{Data})}.$$

This conditional probability presumes that the general proposition G has a probability, which may be logical or subjective. Since

$$\Pr(G) = \Pr(I(G) = 1),$$

some may not agree that the prior probability $\Pr(G)$ can be assigned. But we know $I(G)$ is either zero or one, although unknown. Lee (2020) derived a confidence density for this sunrise problem, giving complete confidence

$$C(E; n \text{ days of consecutive sunrises}) = 1,$$

as well as

$$C(G; n \text{ days of consecutive sunrises}) = 1 \text{ for all } n.$$

Even though $C(E; n$ days of consecutive sunrises) and $C(G; n$ days of consecutive sunrises) correspond to Bayesian posteriors $\Pr(E|n$ days of consecutive sunrises) and $\Pr(G|n$ days of consecutive sunrises), respectively, we don't use the probability notation because confidence is not probability.

In the Bayesian approach, the current posterior $\Pr(G|n$ days of consecutive sunrises) becomes the prior $\Pr(G)$ for future data. Thus, once $\Pr(G|n$ days of consecutive sunrises) $= 1$, $\Pr(G|$no sunrise after n days of consecutive sunrises) cannot be altered. Thus, in the Bayesian approach, a paradigm shift is not allowed. However, $C(G;$ no sunrise after n days of consecutive sunrises) $= 0$. The confidence corresponds to an implied prior of Beta$(1, 0)$, which implies that only one success is observed a priori. Thus, if we observe all successes, it is legitimate to attain 100% confidence in $\theta = 1$. This resolves the induction problem: we can have complete confidence in propositions by means of induction. Furthermore, Lee (2020) showed that confidence gives a new way of hypothesis testing.

Lee also showed that the use of reference prior Beta$(1/2, 1/2)$ cannot avoid a probability dilution by giving the posterior $\Pr(G|T_n = n) = 0$. In Jeffreys's resolution (Section 8.5), $\Pr(G|T_n = n) > 0$, because he assumes $\Pr(G) = 1/2$, but he still cannot reach complete confidence with finite evidence. However, the confidence resolution surprisingly implies that it is legitimate to claim $C(G) = \Pr(G|\text{Data}) = 1$, i.e., complete confidence in the general proposition, even with just one sample. The confidence approach interprets this complete confidence as a consistent, sample-dependent frequentist estimator of the unknown true status $I(G)$ of the general proposition G. The estimator becomes more accurate as the amount of evidence grows. Recently, Lee and Lee (2023a) showed that the confidence distribution can avoid probability dilution even in Stein's (1959) paradox and satellite conjunction paradox.

Let's now illustrate with a great milestone in science. To confirm the validity of the general relativity theory, the observational evidence of light bending was obtained in 1919, and the astrophysical measurement of the gravitational redshift was obtained in 1925. That is, just a few observations were needed to confirm Einstein's new theory with complete confidence. The theory will continue to be used to predict novel phenomena until there is a better theory. We also expect the theory to be uniformly true across time and space. Such inductive reasoning is consistent with scientific theory and, therefore, rational. So, by means of induction based on the evidence so far, we should be able to say with complete confidence that the Sun always rises, until it doesn't. (Of course, from physics, we know that the sun will eventually run out of energy, and the solar system will vanish, but here we are discussing only our logical-mathematical confidence given some evidence.)

To establish complete confidence in a general proposition from particular instances using induction, scientists do not necessarily have to review all instances, but they do need to establish a causal explanatory theory pertaining to the generation of the instances. If one drops an apple, one can be sure that it will fall unless the Newtonian laws suddenly stop working. Indeed, it is induction, as we have seen, that allows such uniformity in scientific theory, and its glory goes to both science and philosophy.

12

Extended Likelihood

Many variants of the likelihood have been introduced to deal with complex data and modelling issues: (i) the nuisance parameter problem, such as the conditional, marginal, estimated, partial and empirical likelihoods, or (ii) the non-Gaussian data, such as the quasi likelihood. All these variants are crucial in the development of likelihood-based statistical models that have grown since the 1970s, particularly generalized linear models (GLMs) and survival analysis; see Pawitan (2004) for more details. But we shall now focus on special extensions of the likelihood to deal with random parameters: extended likelihood and h-likelihood. Surprisingly, confidence turns out to be an extended likelihood; it's surprising since the two concepts appear unrelated. This connection extends the utility of the confidence concept as a measure of uncertainty outside the confidence interval procedure. Last but not least, we use the extended likelihood to resolve the wallet game paradox, which shows that the step from probability to extended likelihood becomes logically necessary once we are dealing with a realized but still unobserved random variable.

12.1 Random Parameters

Fisher's classical likelihood is defined for fixed parameters (Section 10.1), but statistical models involving random parameters are common. Random parameters allow flexible modelling at the individual level. For instance, if we have repeated measures over time, then we can fit a subject-specific regression model, each with its own intercept and slope. The result will be linear mixed models that contain fixed and random effects. Random parameters also naturally arise in prediction and missing-data problems. In the statistical literature, random parameters appear with various names such as random effects, latent processes, factor, missing data, unobserved future observations, potential outcomes, etc. Estimation, sometimes called prediction, of random parameters was traditionally based on minimizing the mean-squared prediction error. This is a natural approach for Gaussian data, but the growth in non-Gaussian data analysis creates the need for a likelihood-based approach.

Given an unknown fixed parameter θ and unknown random parameter u, and source data y, we start with the following:

> **Definition:** The extended likelihood of (θ, u) based on y is the joint probability of (y, u) as a function of (θ, u), i.e.,
>
> $$L_e(\theta, u; y) \equiv p_\theta(y, u). \tag{12.1}$$

By decomposing the joint probability of (y, u) in two different ways, we may write the extended likelihood as

$$\begin{aligned} L_e(\theta, u; y) &= p_\theta(y|u)p_\theta(u) \equiv L(\theta, u; y)L_e(\theta, u) & (12.2) \\ &= p_\theta(y)p_\theta(u|y) \equiv L(\theta, y)L_e(\theta, u; u|y). & (12.3) \end{aligned}$$

In the first decomposition (12.2), conditional on the random parameter u, the data y is generated from a probability model $p_\theta(y|u)$. Here, u behaves like a fixed parameter, so we have the classical likelihood $L(\theta, u; y)$. The random parameter u is assumed to be generated from a model $p_\theta(u)$, so we have an extended likelihood $L_e(\theta, u)$. This decomposition is useful during the construction of the extended likelihood.

The second decomposition (12.3) tells us about information. The first factor $L(\theta, y)$ is the classical likelihood of the fixed parameter θ. According to the likelihood principle (Section 10.4), it contains all the information about θ in the data y, so the second factor $L_e(\theta, u; u|y)$ cannot have additional information about θ. Since the first factor does not involve u, intuitively, the second factor captures the information about u in the data y. Formally, Bjørnstad (1996) proved the extended likelihood principle, which implies all the information about (θ, u) in the data is in the extended likelihood. Together with the classical likelihood principle (Section 10.4), this means the second factor must contain all the information about u in the data.

In the definition (12.1), the random variable u is *not* data. It is only logically connected to the data y, so the extended likelihood is truly distinct from the classical likelihood, which is determined by the probability of the observed data alone. As a quick example, suppose that we have a scalar u sampled from $N(0,1)$, and nothing else. It is important to presume that u is *realized, but still unknown.* If u were known, then there is no uncertainty and the likelihood concept would have no value. An assumed realized u makes it take a specific value, not an abstract random variable. Technically, time makes no difference: A realized future value is also allowed. For instance, we can draw u *tomorrow.* Statistically, as long as the realized value is unknown, there is no difference between yesterday or tomorrow. What is the classical likelihood of u? As we do not have any observations parameterized by u, the classical likelihood of u is the trivial constant likelihood, or $L(u) \equiv 1$. But the extended likelihood of u is non-trivial: $L_e(u) = \phi(u)$, the standard normal density. In general, for any random unknown u generated from a probability model $p(u)$ we have $L_e(u) = p(u)$.

Although extended likelihood had been discussed in the late 1980s (Butler, 1986; Bjørnstad, 1990), its suggested application was fairly limited, such as time series forecasting problems. It was Lee and Nelder (1996) who realized the fundamental importance of the extended likelihood for extending the GLMs into hierarchical GLMs (HGLMs), an extremely rich framework for statistical modelling involving fixed and random parameters.

Why did so many people miss the importance of the extended likelihood? There is a major logical hurdle in the practical use of the extended likelihood: In contrast to the classical likelihood, the extended likelihood lacks invariance with respect to transformations of the random parameter. This is obvious from the definition (12.1). Assuming u is continuous, transforming u will produce a Jacobian term, so the choice of scale matters. The problem is serious: naively treating the extended likelihood as a classical likelihood can easily lead to contradictions. For instance, equivalent models of the random parameter on different scales can lead to different MLEs for the *fixed parameter.* This makes extended likelihood a seemingly useless concept.

As an illustration, consider the one-way random-effect model

$$y_{ij} = \mu + v_i + e_{ij}, \text{ for } i = 1, \ldots, m; \; j = 1, \ldots, n,$$

where $v = (v_1, \ldots, v_m)$ are independent and identically distributed (iid) as $N(0,1)$, and conditionally on v_i, the outcomes y_{ij}s are iid $N(\mu + v_i, 1)$.

The extended log-likelihood of μ and v is

$$\log L_e(\mu, v; y) = -\frac{1}{2}\sum_{ij}(y_{ij} - \mu - v_i)^2 - \frac{1}{2}\sum_i v_i^2.$$

Joint maximization over μ and v gives the standard MLE $\widehat{\mu} = \bar{y}_{..}$, the overall average. However, if we reparameterize v in terms of the log-normal distribution, so that $v_i \equiv \log u_i$, then we can set up another extended likelihood based on $u \equiv (u_1, \ldots, u_m)$, given by

$$\log L_e(\mu, u; y) = -\frac{1}{2}\sum_{ij}(y_{ij} - \mu - \log u_i)^2 - \frac{1}{2}\sum_i(\log u_i)^2 - \sum_i \log u_i.$$

The last term on the right-hand side is the Jacobian term. The two models in terms of v or u are of course equivalent, *but* now a joint maximization over μ and u gives $\widehat{\mu} = \bar{y}_{..} + 1$! Without a general principle to deal with this problem, the extended likelihood concept would indeed be useless.

Lee and Nelder's major contribution was to show that there is a *special scale of the random parameter*, later called the canonical scale, such that the extended likelihood behaves like the classical likelihood and avoids the problem above. They called the special extended likelihood the *h-likelihood.*

12.2 Hierarchical Likelihood*

Since Fisher (1922) introduced the likelihood for statistical models with fixed unknown parameters, it has been widely used for statistical inferences. The maximum likelihood estimates (MLEs) are asymptotically best for any function of parameters of interest, achieving the Cramer-Rao lower bound. Furthermore, the Fisher information and associated delta method provide the necessary standard errors of the MLEs. Thus, many inferential questions can be answered using his MLE method. This makes his method popular in statistical inferences. However, an extension of the MLE method is needed for general models with additional random unknowns (unobservables; Berger and Wolpert, 1984; Butler, 1986). Thus, it is of interest to have an extended likelihood, whose maximization gives the best unbiased predictors (BUPs) for random unknowns in addition to MLEs for fixed unknowns. However, despite many attempts, a proper extended likelihood for maximization remains unknown. Bayarri et al. (1988) summarized the difficulties by showing that various versions of existing extended likelihood cannot give sensible estimators for fixed parameters and predictors for random parameters.

Because of the Jacobian term associated with random unknowns, inferences for fixed and random parameters are very different. Lee and Nelder (1996) proposed the use of the hierarchical (h-)likelihood, defined on a particular scale of random parameters. Their aim was that the maximum h-likelihood estimators (MHLEs) give MLEs for fixed parameters and at the same time asymptotically BUPs for random parameters. Their h-likelihood is an extension of Henderson's (1959) joint likelihood for normal linear mixed models to some general models as their MHLEs provide MLEs for the mean parameters but not for the variance components. The Laplace approximation has been advocated to obtain parameter estimators. However, such an approximation gives only approximate MLEs that could be severely biased, especially in binary data.

Consider a multiplicative model

$$\mu_{ij} = \mathrm{E}(y_{ij}|u_i) = u_i \cdot \exp(\mathbf{x}_{ij}^T\boldsymbol{\beta})$$

to give a linear predictor

$$\eta_{ij} = \log \mu_{ij} = \mathbf{x}_{ij}^T\boldsymbol{\beta} + v_i,$$

where $v_i = v(u_i) = \log u_i$. Now we can have two extended likelihoods,

$$L_e(\boldsymbol{\theta}, \mathbf{u}; \mathbf{y}) \equiv p_{\boldsymbol{\theta}}(\mathbf{y}, \mathbf{u}) \quad \text{and} \quad L_e(\boldsymbol{\theta}, \mathbf{v}; \mathbf{y}) \equiv p_{\boldsymbol{\theta}}(\mathbf{y}, \mathbf{v}).$$

When $\mathbf{u}$ is discrete, they provide the common MHLEs (Lee and Bjørnstad, 2013). However, when $\mathbf{u}$ is continuous, due to the Jacobian term $|d\mathbf{u}/d\mathbf{v}|$,

$$L_e(\boldsymbol{\theta}, \mathbf{v}) = L_e(\boldsymbol{\theta}, \mathbf{u}) \left|\frac{d\mathbf{u}}{d\mathbf{v}}\right| \neq L_e(\boldsymbol{\theta}, \mathbf{u}),$$

they lead to different MHLEs. Lee and Nelder (1996) proposed the use of $L_e(\boldsymbol{\theta}, \mathbf{v})$ as the h-likelihood, where $\mathbf{v}$ is an additive scale to the fixed effects in the linear predictor. However, their MHLEs could not give sensible estimators.

Lee and Lee (2023c) reformulated the h-likelihood as follows: We want to find a scale $\mathbf{u}$ such that

$$H(\boldsymbol{\theta}, \mathbf{v}; \mathbf{y}) \equiv L_e(\boldsymbol{\theta}, \mathbf{v}; \mathbf{y})\ a(\boldsymbol{\theta}; \mathbf{y}) = L_e(\boldsymbol{\theta}, \mathbf{u}; \mathbf{y}) \tag{12.4}$$

enables us to obtain sensible MHLEs, where $a(\boldsymbol{\theta}; \mathbf{y}) = |d\mathbf{v}/d\mathbf{u}|$ is a function of $\boldsymbol{\theta}$ and $\mathbf{y}$. Such a function always exists. For example, let $u_i = p_\theta(\widetilde{v}_i|\mathbf{y}) \cdot v_i$, i.e., $a(\boldsymbol{\theta}; \mathbf{y}) = 1/p_{\boldsymbol{\theta}}(\widetilde{\mathbf{v}}|\mathbf{y})$, where $\widetilde{\mathbf{v}} = \text{argmax}_{\mathbf{v}} p_{\boldsymbol{\theta}}(\mathbf{v}|\mathbf{y})$. Then, we have

$$H(\boldsymbol{\theta}, \widetilde{\mathbf{v}}) = L(\boldsymbol{\theta}; \mathbf{y}) \equiv p_{\boldsymbol{\theta}}(\mathbf{y}).$$

This implies that MHLEs always give MLEs for θ. Furthermore, the resulting h-likelihood satisfies the first two Bartlett identities, so that many of the standard theoretical results from classical likelihood apply to the h-likelihood. Consequently, MHLEs give asymptotically best predictions by achieving the generalized Cramer-Rao lower bound for mixed estimation of fixed and random parameters. Furthermore, the information matrix from the h-likelihood and associated delta method provide the necessary standard errors of the MHLEs. Thus, many inferential questions about both fixed and random parameters can be answered by the MHLE method. For detailed development, see Lee and Lee (2023c).

As an illustration, let's return to the one-way random-effect model. Here, Lee and Nelder's (1996) original h-likelihood is Henderson's (1959) joint likelihood,

$$\log L_e\left(\boldsymbol{\theta}, \mathbf{v}\right) = -\frac{1}{2}\left[\sum_{i,j} \frac{(y_{ij} - \mu - v_i)^2}{\sigma^2} + \frac{\mathbf{v}^T\mathbf{v}}{\lambda^2} + N\log\sigma^2 + q\log\lambda^2\right],$$

where $\boldsymbol{\theta} = (\mu, \sigma^2, \lambda^2)^T$ and $\mathbf{v} = (v_1, ..., v_m)^T$. However, maximization of $L_e\left(\boldsymbol{\theta}, \mathbf{v}\right)$ cannot give the MLEs of variance components σ^2 and λ^2. Now, maximization of the new h-likelihood

$$h(\boldsymbol{\theta}, \mathbf{v}) = \log H(\boldsymbol{\theta}, \mathbf{v}) = \log L_e(\boldsymbol{\theta}, \mathbf{v}) - \frac{q}{2}\log\left(\frac{n\lambda^2 + \sigma^2}{\lambda^2\sigma^2}\right)$$

gives MLEs of $\boldsymbol{\theta}$ while giving BLUPs $E(\mathbf{v}|\mathbf{y})$ of $\mathbf{v}$.

The extended likelihood principle (Bjørnstad, 1996) does not tell us which extended likelihood to use and how to use it for statistical inferences. An extended likelihood with a specific scale of the random parameter gives the h-likelihood, from which optimal MHLEs and their standard errors can be obtained. Thus, Fisher's MLEs for fixed unknowns can be properly extended to general statistical models with additional random unknowns. With the surge of deep neural networks, it is of interest to predict for high-dimensional big data. Lee and Lee (2023a) showed that the MHLE procedure gives the most efficient algorithm of a deep neural network for such datasets. Based on the h-likelihood, we can not only perform a joint maximization of the fixed and random effects, but also perform joint inferences. So, overall, it allows for a unified view and straightforward analyses of HGLMs. Specific modelling issues and examples are beyond the scope of this book; they are covered in Lee, Nelder and Pawitan (2017).

12.3 Confidence Is an Extended Likelihood

Confidence and likelihood are fundamental statistical concepts with distinct technical interpretations and usage. Confidence is a meaningful concept of uncertainty within the context of the confidence interval (CI) procedure, whereas likelihood has been used predominantly as a tool for statistical modelling and inference given observed data. It turns out that there is a much closer connection between the two concepts: Confidence is an extended likelihood.

This result gives the confidence concept an external meaning outside the confidence-interval context, and vice versa; it gives the confidence interpretation to the extended likelihood and provides the extended likelihood direct access to the frequentist probability, an objective certification not directly available to the classical likelihood.

We start by asking if there is a probabilistic way to state our sense of uncertainty in an observed CI. Let T be an estimate of a scalar parameter θ, and θ_0 be the true unknown parameter. Based on T, we construct the $(1-\alpha) \times 100\%$ confidence limits $\widehat{\theta}_\ell$ and $\widehat{\theta}_u$. Define U as the indicator function of whether the CI covers the truth:

$$U \equiv I(\widehat{\theta}_\ell \leq \theta_0 \leq \widehat{\theta}_u),$$

so U is a binary random variable with probability

$$\Pr(U = 1) = 1 - \alpha.$$

For the sake of argument, let's assume that this coverage probability is correct, either exactly or asymptotically. As U is unknown, let's think of it as a random parameter. What can we say about its uncertainty?

Given an estimate $T = t$ of θ, one might think that we should use $\Pr(U = 1|T = t)$ to capture the sense of uncertainty in U. In fact, from the standard prediction theory, we have an optimal predictor of U, which is $E(U|T = t)$ (e.g., Lee, et al. 2017, Section 4.6). Unfortunately, this probabilistic reasoning does not lead us to any new measure or insight, since

$$\begin{aligned} \Pr(U = 1|T = t) &= E(U|T = t) \\ &= U(t) \\ &= I(\widehat{\theta}_\ell \leq \theta_0 \leq \widehat{\theta}_u), \end{aligned}$$

as U is a function of t through $\widehat{\theta}_\ell$ and $\widehat{\theta}_u$. For an observed CI, the value $U(t)$ is realized but still unknown, because the true parameter is unknown. Hence we arrive at an oft-repeated frequentist statement that 'either the CI covers the truth or it does not, no probability statement can be made concerning it.' It is a vacuous tautological statement of no epistemic value, since it is always true, and one could say it without any theory at all. But U is not completely unknown, e.g., we know its expected value is equal to the confidence level.

Given an observed CI, the random parameter U above is now realized, say $U = u$, but still unknown. As with random-effect parameters in general, while we do not know its value, we can talk about its extended likelihood. By definition, $L_e(u) = p(u) = \Pr(U = u)$, so in particular

$$L_e(u = 1) = \Pr(U = 1) = 1 - \alpha.$$

Hence our sense of uncertainty in the observed confidence interval is captured by the extended likelihood.

The connection to extended likelihood suggests a wider utility of the confidence concept to cover uncertainty about any realized random parameter, not just within the CI context. We shall see that sometimes the term 'confidence' sounds more intuitively appealing than 'extended likelihood.' As a simple example, suppose y is the outcome of a fair coin toss. It is realized, but not revealed. What's your uncertainty that y is heads? Intuitively, you'd say 50-50, but in what technical sense? Bayesians would have their subjective probability, but non-Bayesians cannot use probability. The indicator variable

$$U \equiv [Y = \text{heads}]$$

defines a binary random parameter. So, when y is realized but still unknown, we can talk about our confidence in $u = 1$, which is equal to the extended likelihood

$$L_e(u = 1) = \Pr(U = 1) = \Pr(Y = \text{heads}) = 0.5.$$

Thus, technically, our 50-50 sense of uncertainty in the specific toss is captured by confidence.

Now let's consider a general random parameter $V = v$ generated from a known density $p(v)$. As a parameter, v is realized but unknown. We have the probability

$$\Pr(V \in [v_o, v_o + dv]) \approx p(v_o)dv,$$

implying that we have approximately $100 \times p(v_o)dv\%$ confidence in $v \approx v_o$ for any value v_o, so $p(v)$ works as a confidence density as well as an extended likelihood. The term 'random parameter' is used to indicate that v is realized but unknown, but it can start from any random variable. Theoretically random variables are ruled by probability laws, but once they're realized, they're no longer random variables. So it's not surprising that different rules will apply. As the wallet game paradox below shows, once v is realized, the step from probability to confidence becomes logically necessary.

We previously discussed (Section 11.3) that, although confidence is not probability, it has an additivity property in its native parameter space. As an illustration, let's come back to Basu's ticket paradox on page 152 and discuss how the confidence/extended likelihood overcomes the weakness of the classical likelihood in its lack of additivity.

Ticket paradox revisited. In a box containing 1000 tickets: 20 are marked with a number θ and 980 with 10θ. You take a ticket at random and observe the number $y = 15$. You do not know what θ is, but let's put the probability statement in a more suggestive way:

$$\Pr(Y = 10\theta) = \Pr(\theta \in \{Y/10\}) = 0.98.$$

That is, $Y/10$ is the '98% confidence interval' for θ. In other words, once you observe $y = 15$, you have 98% confidence in $\theta = y/10 = 1.5$.

> And this is equivalent to the extended likelihood $L_e(\theta = 1.5) = 0.98$, although the confidence terminology clearly sounds more intuitive here. At this point, we do not yet get any benefit from the confidence concept.
>
> Previously the issue appeared when the 980 tickets were re-labelled using $a_i\theta$, for $i = 1, \ldots, 980$, where a_is are known distinct values between 9.9 and 10.1. Inference based only on the classical likelihood becomes misleading, since it would say that the MLE $\widehat{\theta} = y = 15$ is 20 times more likely than any of the 980 values of θ that cluster tightly around 1.5. The problem is that the classical likelihood is not additive, so clustering does not mean anything. But here, confidence *is additive*; we sketch a proof in this chapter's appendix. So, after observing $y = 15$, we can add up the confidence in all possible values of $\theta \approx 1.5$ and get the total confidence of 0.98 as before.

The confidence concept in general, including the confidence distribution and confidence density in Section 11.3, has a useful interpretation in terms of extended likelihood. In particular, the confidence density can be written in the form

$$c(\theta; y) \equiv c_0(\theta; y)L(\theta; y), \tag{12.5}$$

where $c_0(\theta; y) \propto c(\theta; y)/L(\theta; y)$ will be called the implied prior confidence, and $L(\theta; y)$ is the classical likelihood. The formula represents an updating rule for a prior by the likelihood similar to the Bayesian rule for updating the prior probability (Section 8.4). However, the similarity is only mathematical, since conceptually confidence is not a probability, but an extended likelihood.

Treating confidence as a frequentist measure, in general we cannot attach the confidence value to an observed CI; i.e., confidence is in general not epistemic. However, we often wish to do so because we naturally feel attached to our study results. In fact, that was Fisher's wish with his fiducial probability, which motivated the confidence interval. He rejected Bayesian epistemic probability because of its use of arbitrarily priors, but there was no non-Bayesian epistemic probability. Many paradoxes of the fiducial probability (Section 11.2) stem from the assumption that it is a genuine Kolmogorov probability. Recognizing the fiducial as extended likelihood, which does not follow probability laws, avoids the paradoxes. Do we get something more? Since likelihood quantities refer to the observed data, we might speculate a general statement that, under some conditions, the confidence/likelihood value applies epistemically to the observed CI. We shall discuss this in the next chapter.

12.4 Wallet Game Paradox

Once the random parameter u is realized, it is no longer a random variable; its uncertainty is captured by the extended likelihood, but the extended likelihood is not a probability. Therefore, the concept here is distinct from the Bayesian approach, which would continue to treat u as a random variable with probability distribution. The fact that extended likelihood is not a probability is a key property that we use to explain the so-called wallet game paradox (Gardner, 1982):

> Two people, call them A and B, equally rich or equally poor, meet to compare the contents of their wallets. Each is unaware of the contents of the two wallets. They agree that whoever has less money receives the contents of the other's wallet. Person A can reason, 'I have a fixed amount of x_1 in my wallet; either I lose it or I win an amount of $x_2 > x_1$ with a 50-50 chance. Therefore, the game is favourable to me.' B can of course reason in exactly the same way. So, the game is advantageous to both players, but by symmetry, the game is fair. Where is the mistake in their reasoning?

To avoid trivialities, we should emphasize that, as presented below, it is easy to show that the game is fair. So, the key question is not to show that the game is fair, but to explain what is wrong with the reasoning above. It should be noted that Gardner himself admitted that he could not identify what was wrong with the reasoning. There is a long discussion about the paradox in Wikipedia, which we highly recommend, but will not repeat here.

Let X_1 and X_2 be the random amounts of money in the two wallets. Assuming that A and B are 'equally rich or equally poor', X_1 and X_2 are iid samples from a positive-valued distribution. For simplicity, also assume that they are continuous. We first note that in this game we actually do not observe any data. Secondly, it is obvious from this setup that the game is fair, i.e., on average, each person would receive $E(X_1)$ when they play the game repeatedly. However, this does not explain what's wrong with the reasoning in the paradox.

Now consider the specific realizations $X_1 = x_1$ and $X_2 = x_2$ (but still unknown to both players). As player A, you think $x_1 < x_2$ or $x_1 > x_2$. If the former occurs, you gain x_2, if the latter occurs, you lose x_1. In other words, since x_1 is fixed, you think that if you lose, your loss is limited to x_1. But if you win, your gain is x_2, which by construction is greater than x_1. You are, of course, uncertain whether you will win or lose. If we account for this uncertainty by attaching probability 0.5 to those specific events, then the expected gain minus loss is

$$0.5(x_2 - x_1) > 0,$$

hence the paradox. Only the probability reasoning allows for this final averaging step. Now let

$$U \equiv I(X_1 < X_2),$$

so U is a 0-1 Bernoulli random variable with

$$\Pr(U = 1) = \Pr(X_1 < X_2) = 0.5.$$

However, the specific realization of the current game, either $u = 0$ or 1, is unobserved, so our mind is in a state of uncertainty. Additionally, there is no way to get information about the amount of money in the wallet. If there were, it would be cheating. So, this uncertainty is not caused by random errors and there is no way to change this uncertainty. This shows that the paradox arises from the use of probability to represent the epistemic uncertainty of a single event u.

The question is: What then is this sense of uncertainty about whether the realized u is 0 or 1? It cannot be a probability, since it must then allow for an averaging step that would lead to the paradox. In fact, by definition (12.1), it is given by the extended likelihood, i.e.,

$$\begin{aligned} L_e(u = 0) &\equiv \Pr(U = 0) = 0.5, \\ L_e(u = 1) &\equiv \Pr(U = 1) = 0.5, \end{aligned}$$

so the specific realizations of u are equally likely, not equally probable. We have no theory on how to take an expectation using the extended likelihood as a weight, so the paradox is avoided. The problem for the layman is that expressions such as '50-50 chance', 'equally probable' or 'equally likely' have similar meanings. But technically, different types of uncertainty have different rules.

The wallet paradox shows that the step from the probability of a random event to the extended likelihood becomes logically necessary once we are dealing with a realized (but still unobserved) value.

Unfavourable version

A small variation of the story generates a different paradox, where it becomes unfavourable to play the game. This variation illustrates further the unexpected implications of our views of probability and uncertainty:

> Suppose A opens his wallet and reveals \$100 to both of them. Now B – assumed still completely ignorant of the amount in his own wallet – thinks: 'either I would win \$100, or lose an amount greater than \$100, still with a 50-50 chance, so now it is a losing game for me.' However, again by symmetry, the game is still fair.

This paradox can be explained in a similar way. Now the unobserved random outcome is

$$U = I(X_2 > 100),$$

so we have the extended likelihood

$$L_e(u = 1) = \Pr(U = 1) = 0.5$$

for the realized but unknown u, and the paradox is similarly avoided.

Resolution via utility theory*

Another resolution of the wallet paradox is possible via utility theory (Section 7.4), but we will first re-interpret the theory in terms of confidence-weighted utility. Suppose we bet on one of k possible random outcomes with known probabilities p_i for $i = 1, \ldots, k$, with respective winnings $w_1, \ldots, w_k$. The expected utility of the bet is

$$\mathcal{U} \equiv \sum_i p_i u(w_i), \tag{12.6}$$

where the $u(\cdot)$ is the utility function, which is subjective and commonly assumed to be concave to represent risk aversion. Let's also assume that $u(0) \equiv 0$.

In VNM's rational decision theory, the expected utility is usable in a specific bet, meaning that it's computed for a specific realization of the random outcome. Thus, from the previous discussion, the probability of winning w_i is logically turned into extended likelihood or, more suggestively, confidence:

$$C(w_i) = L_e(w_i) = \Pr(W = w_i) = p_i.$$

In other words, we can use confidence as a weight to calculate the expected utility (12.6), and the expected utility is actually a confidence-weighted utility.

As before, suppose that A has x_1 in his wallet and B has x_2. From A's point of view, his current utility is $u(x_1)$. He has 50-50 confidence of ending up with 0 or with $x_1 + x_2$, so his expected utility is

$$0.5 \times u(x_1 + x_2).$$

Assuming $u(\cdot)$ is concave, even though $x_2 > x_1$, there is no guarantee that

$$0.5 \times u(x_1 + x_2) > u(x_1),$$

so betting is not obviously favourable to him. B will have a similar perspective, so the paradox is avoided. The problem with this solution is that, in small amounts, the utility of money is typically assumed linear, in which case the paradox persists.

Resolution via prospect theory*

Another resolution that allows linear utility is possible via prospect theory (Kahneman and Tversky, 1979), also discussed in Section 14.1. The prospect score of a bet is defined as

$$\mathcal{V}(\text{bet}) = \sum_i \pi(p_i) u(w_i),$$

where p_i, w_i and $u(w_i)$ are the same as for the expected utility above, but now we assume $u(w_i) \equiv w_i$. Furthermore, as for the expected utility, we can interpret p_i as confidence rather than probability. The new element $\pi(\cdot)$ is the decision weight function, having values $\pi(0) \equiv 0$ and $\pi(1) \equiv 1$ and a sub-certainty property $\pi(p) + \pi(1-p) < 1$ for $0 < p < 1$.

Now, from A's perspective, his score if he does not play the game is

$$\mathcal{V}(\text{no play}) = \pi(1) x_1 = x_1.$$

If he plays, his score is

$$\mathcal{V}(\text{play}) = \pi(0.5) \times (x_1 + x_2).$$

The sub-certainty property implies $\pi(0.5) < 0.5$, so even though $x_2 > x_1$ there is no guarantee that

$$\mathcal{V}(\text{play}) = \pi(0.5) \times (x_1 + x_2) > x_1 = \mathcal{V}(\text{no play}).$$

B will have a similar perspective, so the paradox is avoided.

Appendix: Additivity of Confidence

A confidence distribution is usually defined starting with the P-value. Assuming we have a valid distribution function, additivity applies to the parameter used in the P-value. However, for the model used in the ticket paradox, the confidence does not start with the P-value, though we show here that additivity still holds. Using a more general model, suppose the outcome Y is such that

$$\Pr(Y = a_i(\theta)) = p_i, \quad \text{for } i = 1, \ldots, k,$$

where k is a known integer, $a_i(\cdot)$s are known 1-1 functions of θ and p_is are known probabilities that add up to one. First, we can rewrite the probability as

$$\Pr(\theta \in \{a_i^{-1}(Y)\}) = p_i,$$

immediately suggesting that, after observing y, we have $100 \times p_i\%$ confidence in $\theta = a_i^{-1}(y)$. To show additivity, take any two $a_i(\cdot)$ and $a_j(\cdot)$ for $i \neq j$:

$$\begin{aligned}
\Pr(\theta \in \{a_i^{-1}(Y) \text{ or } a_j^{-1}(Y)\}) &= \Pr(\theta = a_i^{-1}(Y) \text{ or } \theta = a_j^{-1}(Y)) \\
&= \Pr(Y = a_i(\theta) \text{ or } Y = a_j(\theta)) \\
&= \Pr(Y = a_i(\theta)) + \Pr(Y = a_j(\theta)) \\
&= \Pr(\theta \in \{a_i^{-1}(Y)\}) + \Pr(\theta \in \{a_j^{-1}(Y)\}).
\end{aligned}$$

The proof can be extended to any number of possible values of θ.

13

Epistemic Confidence

The goal of this chapter is to describe *epistemic confidence*, meaning the sense of confidence in unique single events. For confidence intervals, it is the confidence in the true status of *the observed confidence intervals*. Epistemic confidence is not available – or even denied – in orthodox frequentist inference, as the confidence level is understood to apply to the procedure and not to any particular interval. Thus, frequentist 'confidence' actually has no separate meaning from probability; in particular, it has no epistemic property. Yet, there are obvious practical and psychological needs to think about the uncertainty in an observed interval, for example, for decision making. Traditionally, only Bayesians have no problem stating that their subjective inference is epistemic. Is there a way to make non-Bayesian confidence epistemic?

It is a long-standing question in the philosophy of probability whether there is an objective probability of single events. We specifically exclude repeated events, so the term 'objective' here does not mean 'frequentist' but 'non-subjective,' such that all rational individuals can agree. It is also objective in the sense of existing as an imagined reality (page 68). The logical difficulties, mostly associated with the reference class problem, are reviewed on page 87. Because of these difficulties, frequentists simply reject the existence of such a probability.

As reviewed in Chapter 6, the logical theories as proposed by Keynes, Jaynes or Carnap, and to some extent Popper's propensity theory, aspire to provide the proper foundation for an objective probability of single events. But they were never developed sufficiently to provide a convincing practical methodology. In this chapter, we discuss the solution in terms of epistemic confidence. The use of the confidence concept means that it can be widely accepted by non-Bayesians, while the epistemic element means that it applies to single events.

Our key motivation comes from Fisher's (1958) requirements for probability as a logical measure of uncertainty; see page 83. The third requirement is a novel requirement that stipulates the absence of relevant subsets. Recall from page 89 that R is a relevant subset if it contains information such that

$$\Pr(E|R) \neq \Pr(E)$$

for an event of interest E. According to Fisher, $\Pr(E)$ is epistemic if there is no such R. We shall expand this idea in detail below. The absence of relevant subsets addresses the sticky reference class problem, which underlies the difficulties in setting up an objective probability of single events.

Let's start with a simple example that shows the characteristics of epistemic confidence. It will also show that the concept is actually something we have seen regularly. Let Y be the outcome of a fair coin toss, so that

$$\Pr(Y = y) = 0.5, \text{ for } y = \text{ heads, tails.}$$

By construction, this probability does not have relevant subsets. But we're interested in y that is *realized but still unknown*; so formally it's no longer a random variable. It is a single unique event, whose uncertainty is not a probability. By defining a binary random variable $U \equiv [Y = \text{heads}]$, we have

$$\Pr(U = 1) = 0.5,$$

so once y is realized, we have 50% confidence in $u = 1$ or $y = $ heads. Confidence is materially determined by the amount of information. So, it is not a contradiction that for those who have seen the result of the toss, their confidence in $y = $ heads is either 0% or 100%, but for those who have not seen the result, the confidence is 50%.

Every confidence statement is associated with a probability statement, so it is meaningful to have the following:

> **Definition:** Confidence is epistemic if the associated probability has no relevant subset.

So, for the coin example, we have 50% epistemic confidence. We discussed this coin example previously in Section 12.3, but now we want to emphasize its epistemic aspect, meaning that the 50% confidence applies to the specific result. How do we know that? We could, and people do, rationally and intuitively use the 50% value as a betting price in the specific toss. But, if we are betting, what's the difference with the classical subjective Bayesians? The difference is that the 50% is not decided by personal preferences, but by objective probability, so it can be agreed on by all rational individuals.

By extension of the coin example, all bets in classical games of chance, involving cards, dice, etc., are based on epistemic confidence. Perhaps at this point you could be wondering, so we just replace probability by confidence, what's the big deal? Well, remember that virtually all introductory statistics textbooks explicitly state that the confidence level does not apply to the observed confidence interval. And, technically and philosophically, non-Bayesians do not have objective probability for single events. Epistemic confidence is the objective measure of uncertainty that's attached to single events, where the objectivity is based on a consensus of rational minds. Last but not least, theoretically, not all probabilities can be turned into epistemic confidence.

13.1 Non-epistemic Confidence and the Dutch Book

From now on, our discussion will focus on confidence interval procedures. They are theoretically richer than classical games of chance, as there can be non-trivial relevant subsets that make the confidence non-epistemic. First, generically, given data $Y = y$ of arbitrary size and complexity, a $100\gamma\%$ confidence interval $\mathrm{CI}(y)$ is computed with coverage probability

$$\mathrm{Pr}_\theta(\theta \in \mathrm{CI}(Y)) = \gamma.$$

It is intuitively useful to consider the confidence level γ as a betting price.

Masked integer model. Let $y \equiv (y_1, y_2)$ be an iid sample from a uniform distribution on $\{\theta - 1, \theta, \theta + 1\}$, where the parameter θ is an integer. Let $y_{(1)}$ and $y_{(2)}$ be the minimum and maximum values of y_1 and y_2. We can show that the confidence interval $\mathrm{CI}(y) \equiv [y_{(1)}, y_{(2)}]$ has a coverage probability

$$\mathrm{Pr}_\theta(\theta \in \mathrm{CI}) = 7/9 = 0.78.$$

For example, on observing $y_{(1)} = 3$ and $y_{(2)} = 5$, the interval $[3, 5]$ is formally a 78% CI for θ. But, if we ponder a bit, in this case, we can actually *be sure* that the true $\theta = 4$. So, the probability of $7/9$ is clearly a wrong price for this interval. This is a typical example justifying the frequentist objection to attaching the coverage probability as a sense of confidence in an observed CI.

Here, the range $R \equiv R(y) \equiv y_{(2)} - y_{(1)}$ is relevant. If $R = 2$ we know for sure that θ is equal to the midpoint of the interval, so the CI will always be correct. But if $R = 0$, the CI is equal to the point y_1, and it falls with equal probability at the integers $\{\theta - 1, \theta, \theta + 1\}$. So, for all θ, we have

$$\begin{aligned} \Pr_\theta(\theta \in \text{CI}|R = 2) &= 1 > 7/9 \\ \Pr_\theta(\theta \in \text{CI}|R = 1) &= 1 > 7/9 \\ \Pr_\theta(\theta \in \text{CI}|R = 0) &= 1/3 < 7/9. \end{aligned}$$

Note that the distribution of R is free of θ but R carries information about precision, so R is an ancillary statistic (page 10.4). So, we can see here that ancillary statistics are a source of relevant subsets. The information in the ancillary statistic is captured by conditioning on it. An ancillary statistic is called maximal if there are no further relevant subsets once we condition on the ancillary.

In Section 6.5 we described the Dutch Book argument used by Ramsey and de Finetti to justify the subjective probability. It is based on the coherency axiom that no rational person will devise a betting strategy that is guaranteed to lose money. Crucially, the bets are assumed to be played between two people. To allow many people to bet at the same time, which is a key element of objectivity, the Dutch Book argument is extended to a market-based version in Section 6.6. The argument assumes a more general coherency axiom that no rational person will devise a betting strategy that guarantees an external agent to make a risk-free profit, which amounts to an opportunity cost/loss to the person. This kind of profit is possible if there is a discrepancy between your personal price and the market price that the external agent exploits. Assuming perfect market conditions, e.g., equal access to information by all participants, the agent's profit is an opportunity cost to your ignorance.

In the betting market, intelligent players will use the range information to settle prices at these conditional probabilities. We can be sure, for example, that if $y_1 = 3$ and $y_2 = 5$, the intelligent players will not use 7/9 as the price and will instead use 1.00. So, the information can be used to construct a Dutch Book against anyone who ignores R and unwittingly uses the unconditional coverage. How do we know that there is a relevant subset in this case and that it is the range information R? Moreover, given R, how do we know if there is no further relevant subset? We need a theory for that.

To contrast with the classical Ramsey-de Finetti Dutch Book argument, suppose $y_1 = y_2 = 3$. If, for whatever reasons, you set the price 7/9 for $[\theta \in \text{CI}]$, you are being coherent as long as you set the price 2/9 for $[\theta \notin \text{CI}]$, since the two numbers constitute a valid probability. Coherence means that I cannot make a *risk-free* profit from you based on *this single realization* of y. Even if I know, based on the conditional coverage, that 1/3 is a better price, I cannot take advantage of you because there is no betting market. So 7/9 is a valid subjective probability.

13.2 Epistemic Confidence Theorem

Our question is, under what circumstances does the confidence, as measured by the coverage probability, apply to the observed interval? Intuitively, we could use the coverage probability γ as a betting price if there is no better price given the available data. So, the question is, are there any features of the data that can be used to improve the price? Mathematically, these features are some observed statistics that can be used to help predict the true coverage status. Given an arbitrary statistic $S(y)$, the conditional coverage probability $\Pr_\theta(\theta \in \text{CI}|S(y))$ will in general be biased, i.e., different from the marginal coverage. However, the bias as a function of the unknown θ is generally not going to be consistently in one direction. In terms of betting, this means that we cannot exploit an arbitrary feature of the data as a way to construct a Dutch Book against someone who sets the price at γ.

A statistic $R(y)$ is defined as relevant if the conditional coverage is *non-trivially and consistently biased in one direction.* That is, for a positive bias, there is $\epsilon > 0$ free of θ, such that

$$\Pr_\theta(\theta \in \text{CI}(Y)|R(y)) \geq \gamma + \epsilon \quad \text{for all } \theta. \tag{13.1}$$

Now, the feature $R(y)$ *can be used* to construct a Dutch Book: Suppose that you and I are betting and I notice that $R(y)$ is observed. If you ignore it and set the price at γ, then I would buy the bet from you and then sell it on the betting market at $\gamma + \epsilon$. So I make a risk-free profit of ϵ. (We have assumed that the market contains intelligent players, so they would also have noticed the relevant statistic and set the price accordingly.) The negative bias is defined similarly.

Technically, $R(y)$ induces subsets of the sample space; they are the relevant subsets. Therefore, if there is a relevant subset, the confidence level γ is not epistemic. Conversely, if there are no relevant subsets, the betting price determined by the confidence level is protected from the Dutch Book. Therefore, for confidence interval procedures, we arrive at the same definition of epistemic confidence that its associated coverage probability has no relevant subsets.

In the previous chapter, we discuss that confidence is an extended likelihood. The classical likelihood principle and its extended version state that the likelihood contains all the information in the data. Intuitively, this implies that the likelihood leaves no relevant subset and is thus protected from the Dutch Book. In other words, confidence is epistemic if it is associated with full likelihood. The theory is somewhat complex and requires the confidence density from Section 11.3; exact technical conditions and statements are given in Pawitan et al. (2023). The steps can be summarized as follows:

1. Find a statistic $T(y)$ that leads to a valid confidence distribution $c(\theta; t)$. Here $T(y)$ can be any convenient estimate, even from a subset of the data.

2. Compute the implied prior $c_0(\theta; t)$ defined by

$$c_0(\theta; t) \propto \frac{c(\theta; t)}{L(\theta; t)}, \tag{13.2}$$

   where $L(\theta; t)$ is the likelihood based on t alone, with the understanding that $c_0(\theta; t)$ can ignore all terms free of θ.

3. Define the full confidence density as

$$c_f(\theta; y) \equiv c_0(\theta; t) L(\theta; y),$$

   where $L(\theta; y)$ is the likelihood based on the full data y.

Then the confidence statements made from $c_f(\theta; y)$ are epistemic. If the maximum likelihood estimate (MLE) is sufficient, then there is no relevant subset, so the epistemic property is expected. If the MLE is not sufficient and there is a unique maximal ancillary statistic, then the full confidence is automatically conditional on the maximal ancillary, so there are no further relevant subsets. It is one of Fisher's (many) great statistical realizations that conditioning on maximal ancillary recovers the loss of information in a non-sufficient MLE. These rather abstract results can be better understood in the examples below. Note from the examples that, very conveniently, sometimes there is no need to have an explicit formula for the MLE or the maximal ancillary.

In logic and philosophy, the relevant subset problem is known as the reference class problem, which we described at length in Section 6.4. This problem is solved here by setting a limit on the amount of information, as given by the observed data y generated by the probability model $p_\theta(y)$. This approach is in line with the theory of market equilibrium, where the perfect market hypothesis assumes that information is available to all players. It is not possible to have an equilibrium – hence accepted and settled prices – if information is indefinite or when different market participants know that they have access to different pieces of information.

Examples

In many applications where we have independent data $y \equiv (y_1, \ldots, y_n)$, it is simplest to start with just y_1 for the statistic $T(y)$ in Step 1. To illustrate with the masked integer model in the previous section, based on $T(y) = y_1$ alone, the confidence density and the likelihood functions are proportional:

$$c(\theta; y_1) \propto L(\theta; y_1) = 1, \text{ for } \theta \in \{y_1 - 1, y_1, y_1 + 1\},$$

so the implied prior $c_0(\theta) = 1$ for all θ. The full likelihood based on (y_1, y_2) is

$$L(\theta) = 1, \text{ for } \theta \in [y_{(2)} - 1, y_{(1)} + 1],$$

so the full confidence density is $c_f(\theta) \propto L(\theta)$. For example, if $y_1 = 3$ and $y_2 = 5$, we have 100% confidence that $\theta = 4$. And if $y_1 = y_2 = 3$, we only have 33.3% confidence for $\theta = 4$, although we have 100% confidence for $\theta \in \{2, 3, 4\}$. This recapitulates the results using range as the relevant statistic, but *without* having to know it in advance. Also, we now know that there is no other relevant subset.

By far the most important practical application of the epistemic confidence concept is in large-sample datasets where an approximate normal distribution holds. In this case, the confidence density matches the full likelihood function, so the confidence is epistemic. Theoretical examples are given in Pawitan et al. (2023), but here we discuss an illustrative numerical example.

Let the data y_1, y_2, y_3 be iid samples from $N(\theta, 1)$. Because each y_i is symmetric around θ, a CI based on the range $y_{(1)} < \theta < y_{(3)}$ has 75% confidence level.[1] However, the confidence is not epistemic because it is not based on full likelihood, so the reported confidence level does not apply to the observed interval. The standard normal-based CI is epistemic because it is based on the sample average, which is sufficient. To be concrete, let's analyse these two simple datasets

A (narrow range): $0.1, 0.3, 0.5$
B (wide range): $-1.7, 0.3, 2.3$.

Using only range information, the narrow- and wide-range CIs $0.1 < \theta < 0.5$ and $-1.7 < \theta < 2.3$ have 75% confidence. But according to the normal theory, they should have 27.1% and 99.9% epistemic confidence, respectively. These can be computed from the confidence density for the normal model shown in Figure 13.1.

It seems too simplistic to use only range information for inference, but one might feel that the normal theory is not robust. Although the two datasets have different ranges, they have the same average (0.3), hence the same normal-based 75% CI:

$$0.3 \pm 1.15/\sqrt{3} = (-0.36 < \theta < 0.96),$$

thus ignoring the range information. (We choose 75% confidence to allow direct comparisons between all models including the range.)

To account for different ranges in the data without having to model the variability explicitly, let's suppose that the data $y \equiv (y_1, y_2, y_3)$ are iid samples from the Cauchy distribution with location θ and scale 1. In Step 1 above, choose the statistic $T(y) = y_1$, which gives a valid confidence density

$$c(\theta; y_1) \propto \frac{1}{1 + (y_1 - \theta)^2}.$$

[1]To prove this, count $X =$ the number of y_is less than θ. Due to symmetry around θ, X is binomial with $n = 3$ and success probability 0.5. The CI is correct if $X = 1$ or 2, which has the probability 0.75.

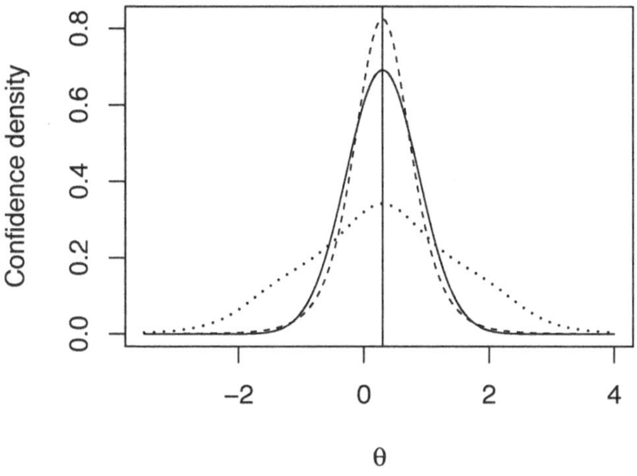

FIGURE 13.1
Confidence densities for the normal model with mean θ (solid curve) and Cauchy models with location θ (dashed and dotted curves for datasets A and B). Dataset A has a short range, while B has a wide range. For the normal model, the two datasets produce the same confidence density. In all cases the MLEs of θ coincide at 0.3 (vertical line).

This is proportional to the likelihood $L(\theta; y_1)$, so the implied prior is constant. Therefore, the complete confidence density for this location model is

$$c(\theta; y) \propto L(\theta; y) = \prod_{i=1}^{3} \frac{1}{1 + (y_i - \theta)^2}.$$

The confidence densities for datasets A and B are shown in Figure 13.1. Under the Cauchy model, the sample average is no longer sufficient; the normal-based 75% CIs have a confidence of 0.79 and 0.41, respectively, for the two datasets. For the second data set, the normal-based CI is too narrow, or the claimed 75% confidence is too optimistic, considering the wide range of data values. In contrast, the Cauchy-based 75% CIs are $-0.30 < \theta < 0.90$ and $-1.18 < \theta < 1.78$, respectively.

In practice, you might be tempted to use a more flexible normal model with unknown variance. It will lead to the standard t-based 75% CIs $0.11 < \theta < 0.49$ and $-1.55 < \theta < 2.15$. These are very close to the range-based CIs. Therefore, both Cauchy- and t-based CIs adapt to the range in the data, with Cauchy-based CIs showing a more moderate adaptation.

Invariance property of the implied prior*

Recall that in Section 9.3 we define invariant priors to have the property that

$$f(\phi) = f(\theta)|J|, \tag{13.3}$$

for 1-1 functions $\phi = h(\theta)$ of the parameter θ, where $|J| = |\partial\theta/\partial\phi|$ is the Jacobian term. With invariant priors, we get consistent inferences regardless of how we choose to parameterize the problem. The implied prior turns out to have the same invariant property.

In the one-parameter case, because 1-1 transformations of the parameter follow the standard rule of probability, the confidence density of ϕ also satisfies:

$$c(\phi; y) = c(\theta; y)|J|.$$

From the definition (13.2), this means that the implied prior of ϕ also follows the same rule,

$$c_0(\phi; y) = c_0(\theta; y)|J|,$$

hence the implied prior is invariant with respect to 1-1 transformations of the parameter.

13.3 Epistemic Confidence and Objective Bayesianism

The epistemic confidence density $c(\theta; y)$ and the implied prior $c_0(\theta; y)$ correspond to the Bayesian posterior $p(\theta|y)$ and the prior $p_0(\theta)$, respectively. That is, we may obtain confidence by using the Bayesian update of the implied prior confidence $c_0(\theta; y)$. Thus, epistemic confidence $c(\theta; y)$ can be considered a non-Bayesian alternative to the Bayesian posterior. However, unlike Bayesian posteriors, epistemic confidence is derived without presuming any explicit prior. The implied prior is implicit in the definition of the P-value and the choice of model. It should also be noted that epistemic confidence is not a frequentist probability. In terms of similarity, priors used in both the objective Bayesian approach and the epistemic confidence method are often improper.

As in the Bayesian posteriors, confidence densities contain a wealth of information for constructing any type of frequentist inference on a fixed parameter. With all the information in the data and the model, the extended likelihood is independent of the state of the individual mind. When we define the P-value to derive confidence, we often use sufficient statistics for θ. If we agree on which statistics to use, then the consequently derived confidence is what a rational man would use. Thus, we may view the implied prior as a prior of the objective Bayesians. This means that epistemic confidence could be the posterior of objective Bayesians and that its epistemic interpretation would be natural to them. However, the implied prior can differ from an existing prior in objective Bayesians as in the sunrise problem.

The confidence distribution for normal means is the same as the Bayesian posterior under the uniform prior. The marginal posterior of the length is equivalent to the integrated confidence density. Stein's (1959) paradox shows that both have severe probability dilution. To solve this problem, Bernardo (1979) proposed the use of a reference prior, but it cannot completely eliminate probability dilution. Lee and Lee (2023b) showed that the confidence distribution solves this problem. This is also very important in the satellite conjunction paradox because poor data often dilute the collision probability. Integrated confidences and marginal posteriors under uniform and reference priors lead to a severe underestimation of the collision probability. According to Lee and Lee (2023b), the confidence distribution provides the appropriate P-values for the hypothesis testing of the collision.

How do subjective Bayesians deal with objective probability? Extra principles are needed to deal with a potential mismatch between subjective and objective probabilities. For instance, what is your *subjective* probability of heads in a specific toss of a coin? If it is 0.5, where do you get the number? Lewis's (1980) 'Principal Principle' states

$$\text{Pr}_s\{A|\text{Chance}(A) = x\} = x,$$

where Pr_s denotes the subjective probability and Chance the objective probability. Our Dutch Book argument can be used to justify the Principle, so the principle does not have to come out of the blue with no logical motivation. However, it should be noted that epistemic confidence is *not simply set equal to probability*. Instead, it is the consequence of a theorem that establishes no relevant subset in order to avoid the Dutch Book. In the setup, the frequentist probability applies to a market that involves a large number of independent players. Moreover, the rational personal betting price is no longer subjective, for example, in the choice of the prior. Thus, the conceptual separation of personal and market prices allows both an epistemic and a frequentist meaning of confidence.

13.4 Probability, Likelihood, Extended Likelihood, H-likelihood and Confidence

We have seen different kinds of uncertainty: classical likelihood, confidence, subjective probability, objective probability and extended likelihood. Classical likelihood is for fixed unknowns, whereas confidence is for a binary random unobservable. We have shown that the confidence is indeed an extended likelihood, which can be used to derive a consensus probability.

According to the Ramsey-de Finetti theorem (Section 6.5), a betting strategy is coherent as long as it satisfies the probability laws. But their definition of probability was subjective. In the confidence approach, epistemic confidence needs an objective scientific theory or a statistical model. Thus, in this sense, the Bayesian subjective probability is wider than the confidence, because a subjective probability does not require a scientific theory for generating the unobservables. However, epistemic confidence refers to the realization of a single event, so confidence can be applied more widely than frequentist probability.

Pawitan et al. (2023) showed that, under appropriate conditions,

$$C(\theta \in \mathrm{CI}_p(y)) = \mathrm{Pr}_\theta(\theta \in \mathrm{CI}_p(Y)),$$

where the former is the confidence of the observed CI for procedure p, and the latter is the coverage probability of the CI procedure p. Lee and Lee (2023b) showed that in Stein's (1959) paradox and satellite conjunction paradox, the coverage probability of an observed CI can be ambiguous, because there are many CI procedures with different coverage probabilities having the same observed CI for a given data y. Due to the severe dilution of probability, probability-based methods cannot produce CIs that can maintain the stated coverage probability. They further showed that the confidence distribution constructed as in Section 11.3 is the only method that produces CI procedures that achieve the stated coverage probability. Based on the confidence distribution, suppose that $\mathrm{CI}_i(Y)$ are frequentist CI procedures for $i = 1, 2, ...$ with confidence level α_i such that

$$\Pr_\theta(\theta \in \mathrm{CI}_i(Y)) = \alpha_i$$

with $\alpha_i \neq \alpha_j$ for $i \neq j$. Lee and Lee (2023b) showed an example that for given data y,

$$\mathrm{CI}(y) = \mathrm{CI}_1(y) = \mathrm{CI}_2(y) = \cdots$$

Thus, the coverage probability of an observed $\mathrm{CI}(y)$ can be ambiguous. They further showed that

$$C(\theta \in \mathrm{CI}(y)) = \max_i \alpha_i = \max_i \Pr_\theta(\theta \in \mathrm{CI}_i(Y)),$$

i.e., the confidence represents the maximum possible coverage probability among the CI procedures producing the given observed CI. Thus, confidence helps us to understand the property of an observed CI which frequentist theory may not cover.

Even after 100 years since its introduction, there is still no general consensus in statistics on the direct use of likelihood for inference, indicating that it is difficult to give a normative answer for it. In the statistical literature, we can point to Edwards (1992) and Royall (1997) as proponents of direct likelihood inference, although we believe that they do not represent the Fisherian views. Fisher recognized two logical levels of uncertainty, one based on probability (fiducial probability, which corresponds to confidence according to our interpretation) and the other based on classical likelihood. They are not meant to be in competition with each other, since the classical likelihood is weaker than the probability.

In the satellite conjunction paradox, probability dilution is a counterintuitive phenomenon that lower-quality data paradoxically appear to dilute the risk of impending collision. This causes a severe and persistent underestimate of risk exposure. Balch et al. (2019) noted probability dilution as a symptom of fundamental deficiency of probabilistic inference. Lee and Lee (2023b) showed that even the marginal posterior based on a reference prior cannot avoid such a probability dilution and the key to preventing it is the use of confidence, i.e., extended likelihood, not probability. The current definition of confidence distribution requires the confidence feature that

$$\Pr(\theta \in \mathrm{CI}(Y)) = C(\theta \in \mathrm{CI}(y))$$

for all θ. In the sunrise problem, probability-based inference, such as the Bayesian posterior, cannot satisfy the confidence feature at $\theta = 1$, whereas in Stein's paradox and the satellite conjunction paradox it fails at $\theta \geq 0$, especially near $\theta = 0$. A point mass at the boundary of parameter space in the confidence distribution helps avoid such a deficiency of probability (Lee and Lee, 2023b). By distinguishing each of these concepts, we can further enrich our understanding of inductive reasoning.

Part IV

Puzzles and Paradoxes

14

Paradoxes of Savage's Axioms

In this chapter, we discuss Allais's and Ellsberg's paradoxes, two well-known paradoxes associated with Savage's axioms (Chapter 7), particularly the sure-thing principle. The paradoxes highlight the descriptive/predictive vs normative roles of the axiomatic system. In its normative role, the axioms act as a guide to human decision making, where a violation would imply that it is the human decision that needs to be corrected to follow the axioms. However, in the descriptive role, the axioms should be able to predict human decisions, so a common violation of any of the axioms raises the question whether the system is appropriate or rich enough to capture real human behaviour. That's how we react to a violation of Euclid's axioms, for instance, in spherical or other non-Euclidean spaces.

14.1 Allais's Paradox

The French economist and Nobel laureate Maurice Allais (1953) described two situations where people tend to prefer alternatives that contradict the expected utility theory (page 113), hence violating at least one of the axioms. It is an important paradox, because among the violators, when Allais first presented it in a 1952 meeting on the economic theory of risk, were future Nobel prize winners in economics – Paul Samuelson, Milton Friedman and Kenneth Arrow. Savage also happened to be there; his reaction will be reported below. The paradox involves the preference order between two bets in two different situations:

Situation 1. Which do you prefer between the following two bets? It's important to use these large amounts of money, because utility is magnitude-dependent, though one can modify \$2,500,000 with smaller amounts such as \$1,000,000 and preserve the paradox:

I.	Win \$500,000 with probability 1,
II.	Win \$2,500,000 with probability 0.1, or \$500,000 with probability 0.89 or nothing with probability 0.01.

Situation 2. Similarly, which of these two bets do you prefer?

III.	Win \$ 500,000 with probability 0.11, or nothing with probability 0.89.
IV.	Win \$2,500,000 with probability 0.1, or nothing with probability 0.9.

If the utility function is linear and we set the baseline $\mathcal{U}(0) \equiv 0$, the expected utility for bet I is $\mathcal{U}(\text{I}) = 500,000$, while for bet II it is

$$\mathcal{U}(\text{II}) = 0.1 \times 2,500,000 + 0.89 \times 500,000 = 695,000.$$

Since $\mathcal{U}(\text{II}) > \mathcal{U}(I)$, you should prefer II over I. But in reality, most people, including legendary economists, would be happy with the smaller but sure amount of money in bet I. The small potential of getting nothing in bet II is not compensated for by the small potential of the larger jackpot. So, people tend to prefer I over II. This is a well-known risk aversion, but wait, it is not the paradox in question yet.

What about between III and IV? A small reduction in the probability of winning \$500,000 in III is rewarded by a large increase in the jackpot in IV. As expected, most people would prefer IV over III. *Now we have the paradox*: that choice turns out to violate the expected utility theory.

To prove it, let's first put the four bets in terms of their expected utilities for a general non-linear utility function $u(\cdot)$ (for convenience, let's drop the 000s from the numbers):

$$\begin{aligned}
\mathcal{U}(\text{I}) &= u(500) \\
\mathcal{U}(\text{II}) &= 0.1u(2500) + 0.89u(500) \\
\mathcal{U}(\text{III}) &= 0.11u(500) \\
\mathcal{U}(\text{IV}) &= 0.1u(2500)
\end{aligned}$$

Preferring I over II implies $u(500) > 0.1u(2500) + 0.89u(500)$, or

$$0.11u(500) > 0.1u(2500).$$

But that means you should prefer III over IV, which contradicts the common preference of people for IV over III.

Kahneman and Tversky (1979) conducted many surveys of people's preferences in situations similar to Allais's and confirmed the offending preferences. From the proof we see that the paradox has nothing to do with non-linear utility or risk aversion. The common explanation is to see the paradox as a violation of the sure-thing principle, Savage's Axiom 2 in Section 7.1.

Violation of Axiom 2

Imagine 100 lottery tickets numbered 1 to 100, where you pick one ticket and a winning number is chosen randomly. The payoffs are given in Table 14.1. Then the four bets in the table are exactly equivalent as above, and we can clearly see that the bets follow the setup of the sure-thing principle. Under the principle, the last column (tickets 12-100) represents an irrelevant event, so a preference for I over II must be consistent with a preference of III over IV.

TABLE 14.1
Allais's four bets are equivalent to the bets in this table, where the columns indicate ticket numbers, and the entries are the payoffs. You pick a number between 1 and 100, and the winning number is selected at random. So, the probability of winning is 0.01, 0.10, and 0.89 for the three columns, respectively.

Bet	1	2–11	12–100
I	500	500	500
II	0	2500	500
III	500	500	0
IV	0	2500	0

Savage initially chose as most people do (i.e., preferring I over II and IV over III), but after realizing that he violated the sure-thing principle, he changed his preference to III over IV. He wrote (Savage, 1972, page 103) 'in reversing my preference between Gambles III and IV I have corrected an error.' But why did he reverse III and IV, not I and II? The sure-thing principle only states that the two pairs of bets must be consistent, but does not say which one is the right choice. In any case, Savage's correction highlights the normative value of the axioms as a guide for rational decision, with the power to correct our subjective preferences. It raises an interesting question whenever there is a conflict: Which has the primacy? Do you modify your preferences or the axioms?

Violation of Axiom 6

The explanation of the Allais paradox as a violation of the sure-thing principle is well known, but we can also argue that the paradox violates the small-event continuity axiom, Savage's Axiom 6 in Section 7.1. Viewed in this perspective, the paradox occurs because of a conflict between Axiom 6 and risk aversion. This is a more serious issue. Let's modify the payoff of some bets on events with small probability, i.e., on ticket 1 with probability 0.01, to get I_0 and III_0. These modifications are shown in Table 14.2.

TABLE 14.2
Modified bets I_0 and III_0 in Allais's paradox, following Table 14.1, where the payoff of some tickets is set to zero. For clarity, the original bets are also listed.

Bet	1	2–11	12–100
I	500	500	500
I_0	0	500	500
II	0	2500	500
III	500	500	0
III_0	0	500	0
IV	0	2500	0

The reasoning follows these steps:

- By the strong-dominance principle (Axiom 7), we must have II $>$ I_0 and IV $>$ III_0.
- III is a small-event modification of III_0 using a payoff of comparable size, so by Axiom 6, we should have IV>III. This is the commonly observed preference; in fact, it is justified by Axioms 6 and 7.
- Now, the modification $I_0 \rightarrow$ I is exactly the same as $III_0 \rightarrow$ III, so again by Axiom 6, we should have II>I. However, the small-event modification from I_0 to I generates *a sure gain*, so for most people the ordering II $>$ I_0 gets reversed to I >II, thus violating Axiom 6.

In this analysis, the preference IV>III is seen not as an error, but a choice that agrees – at least in spirit if not formally – with Axioms 6 and 7. So, as we noted above, there is actually no rational basis for Savage to reverse his initial preference of IV>III. The reason he reversed it was because he first preferred I>II, which was justified by risk aversion. But risk aversion is an empirical phenomenon that is not implied by any of the axioms and plays no role in Savage's theory. Therefore, normatively, reversing the risk-averse preference I>II is perhaps a more consistent choice. But then you would sacrifice risk aversion in order to follow Axiom 6.

Instead of using Axiom 6, we could also invoke the sure-thing principle to justify II> I, but here we want to highlight the violation of Axiom 6. We could also say this: Axiom 6 and the sure-thing principle together contradict risk aversion. Or, risk aversion and the sure-thing principle together contradict Axiom 6. So, to the extent that the small event in Axiom 6 is an approximation of real events with a small probability, the axiom is in conflict with risk aversion. It is a serious conflict, because risk aversion is closely connected to the universally accepted law of diminishing returns.

In view of the discussion following Axiom 6, the reversal of preference occurs even when there is no event with infinitely better or infinitely worse consequences. The reversal – marking a discontinuity – occurs when even a small-event modification creates certainty, as people behave differently when dealing with sure events.

Explanation using the prospect theory

Preferential discontinuity near the boundary of certainty or impossibility is not a rare or exotic phenomenon. As shown by Kahneman and Tversky (1979) it is commonly seen, for example, in gambling and insurance decisions. They took the Allais paradox seriously and developed an alternative axiomatic system called prospect theory, in part to allow for discontinuity when the probability is near 0 or near 1. Instead of maximizing the expected utility, in the prospect theory, a decision maker maximizes the score

$$\mathcal{V} \equiv \sum_i \pi(p_i)u(w_i), \tag{14.1}$$

where $u(w_i)$ is a function of the payoff/winning w_i (they called it a 'value function,' but in principle it works like a utility function), p_i is the probability of state i, and $\pi(\cdot)$ is a decision weight function with $\pi(0) \equiv 0$ and $\pi(1) \equiv 1$. In the classical expected utility theory, $\pi(p_i) = p_i$, but in general, $\pi(p) \neq p$.

An important feature of the theory is that the decision weight function has a discontinuity property: $\pi(0^+) > 0$ and $\pi(1^-) < 1$, and a sub-certainty property $\pi(p) + \pi(1-p) < 1$ when $0 < p < 1$. Furthermore, the prospect score (14.1) can be rewritten as

$$\mathcal{V} = \sum_i p_i \{\pi(p_i)u(w_i)/p_i\} \equiv \sum_i p_i u^*(w_i, p_i),$$

where $u^*(w_i, p_i) \equiv \pi(p_i)u(w_i)/p_i$ becomes a chance-dependent or state-dependent utility function, thus violating Savage's Axiom 3.

Prospect theory is considered one of the cornerstones of behaviour economics; Kahneman was awarded the Nobel Prize in economics in 2002 for this work. In particular, the theory provides better explanations of distinct human reactions to gains and losses, and of behaviour near impossibility or near certainty. Allais's paradox can be explained due to the special behaviour associated with sure gain in Bet I. The scores are

$$\begin{aligned}
\mathcal{V}(\text{I}) &= u(500) \\
\mathcal{V}(\text{II}) &= \pi(0.1)u(2500) + \pi(0.89)u(500) \\
\mathcal{V}(\text{III}) &= \pi(0.11)u(500), \\
\mathcal{V}(\text{IV}) &= \pi(0.1)u(2500).
\end{aligned}$$

The preferences I>II and IV>III imply:

$$\begin{aligned}
\{1 - \pi(0.89)\}u(500) &> \pi(0.1)u(2500) \\
\pi(0.1)u(2500) &> \pi(0.11)u(500).
\end{aligned}$$

A small algebra after adding these two inequalities gives $\pi(0.89) + \pi(0.11) < 1$, which is the sub-certainty property anticipated (and allowed) in prospect theory.

Explanation in terms of simultaneous bets

We close this section with another explanation of the paradox, highlighting an unstated assumption in the setup of the sure-thing principle. Let us assume that you have to make *two simultaneous bets*: (I vs II) and (III vs IV) *based on a single draw of the lottery* in Table 14.3.

TABLE 14.3
Allais's paradox explained in terms of two simultaneous bets on the outcome of a single draw of the lottery in Table 14.1. Preferring I>II and IV>III implies (I+IV)>(II+III), which is irrational because they have exactly the same payoffs.

Bet	1	2–11	12–100
I+IV	500	3000	500
II+III	500	3000	500
I+III	1000	1000	500
II+IV	0	5000	500

Preferring I>II and IV>III implies (I+IV)>(II+III), which is irrational because they have exactly the same payoffs. The preference for one of the other bets (I+III) vs (II+IV) depends on one's utility of money, so there is no immediate inconsistency issue. This means that for the sure-thing principle to hold, the bets are presumed to be made simultaneously based on a single state of the world, e.g., a single draw of a lottery. The presentation of Allais's paradox does not make this assumption. Violators of the principle are likely thinking of two independent bets based on two random draws of the lottery. Had the dependence on the single draw of the lottery been made explicit, a rational person would not likely violate the principle.

14.2 Ellsberg's Paradox

Ellsberg (1961) introduced a thought-provoking paradox that sheds light on the contrasting behaviours exhibited by individuals when faced with ambiguity rather than objective probability. In Donald Rumsfeld's poetic terminology, it's unknown unknown vs known unknown. Economists call the former 'uncertainty' and the latter 'risk,' but we will not do that here, as we have used 'uncertainty' as an omnibus term covering all types of uncertainty. It's an important paradox because it challenges Savage's sure-thing principle and the Bayesian view on uncertainty, which leads to neutrality towards ambiguity.

The paradox involves an urn containing 90 balls, 30 of which are red, and the remaining 60 a mixture of black and yellow of unknown proportion. You are asked to pick one ball from the urn and given the option of

(I) getting $100 if the ball is red, or
(II) getting $100 if the ball is black.

Which option would you prefer? These bets are represented in Table 7.2.

TABLE 14.4
The first betting scenario from Ellsberg (1961): An urn contains 30 red balls and 60 black and yellow balls together, but in an unknown proportion. You pick one ball from the urn. The table shows the payoffs based on the colour of the ball. Do you prefer option I or II?

Option	Red	Black	Yellow
I	100	0	0
II	0	100	0

Now consider the second scenario, represented in Table 14.5: Again, you pick one ball at random from the urn and are given the option of (III) getting $100 if the ball is red or yellow; or (IV) getting $100 if the ball is black or yellow. Which option would you prefer now?

TABLE 14.5
The second betting scenario from Ellsberg (1961): Similar setup as in Table 14.4, but with different payoffs. Do you prefer option III or IV?

Option	Red	Black	Yellow
III	100	0	100
IV	0	100	100

If you are like most people (see, e.g., Camerer and Weber, 1992), you'd prefer I $>$ II and IV $>$ III. It shows that people tend to prefer objective probability over subjective probability. Behavioural economists call this tendency *ambiguity aversion*. This is clearly in violation of the sure-thing principle: Because Yellow carries the same payoffs within each pair of bets, the presence of yellow balls should not influence the preference between red and black balls. Thus, according to the principle, preferring I > II implies preferring III > IV, and vice versa.

Violating the axiom will, of course, lead to a violation of the expected utility theory. Let the unknown proportions of the black and yellow balls be p_b and p_y, respectively, thus we only know that $p_b + p_y = 2/3$. Using a generic utility function $u(\cdot)$ for the winning, the expected utilities of the options are

$$\begin{aligned}\mathcal{U}(\text{I}) &= \frac{1}{3}u(100) \\ \mathcal{U}(\text{II}) &= p_b u(100) \\ \mathcal{U}(\text{III}) &= \frac{1}{3}u(100) + p_y u(100) \\ \mathcal{U}(\text{IV}) &= \frac{2}{3}u(100).\end{aligned}$$

Preferring I > II implies $\mathcal{U}(\text{I}) > \mathcal{U}(\text{II})$, which means

$$\frac{1}{3}u(100) > p_b u(100), \text{ or } p_b < \frac{1}{3},$$

while preferring IV > III implies $\mathcal{U}(\text{IV}) > \mathcal{U}(\text{III})$, which means

$$\frac{2}{3}u(100) > \frac{1}{3}u(100) + p_y u(100), \text{ or } p_y < \frac{1}{3},$$

which together imply $p_b + p_y < 2/3$, a contradiction.

The probability of a black or yellow ball is between 0 and 2/3, but the exact value is unknown. In the Ellsberg paradox literature, it is called an ambiguous probability. Statisticians would simply call it an unknown parameter. The sure-thing principle does not seem to tell us in which direction the preference should be. However, in view of Sinn's (1980) theorem in Section 9.1, together with the weak ordering axiom, the sure-thing principle implies that the ambiguous (uniform) probability representing complete ignorance must be treated like an objective (uniform) probability. This will lead to neutrality towards ambiguity.

To see this, we could use the uniform distribution directly on the number of black balls, so the probability of k black balls is 1/61 for $k = 0, \ldots, 60$. Alternatively, imagine an auxiliary experiment as in Bayes's *Essay* (Section 8.1), where we first perform a binomial experiment to decide the number of black balls, with $n = 60$ trials and an unknown probability.

Let's assume, as Bayes and Laplace did, that the probability of picking a black ball follows the uniform distribution on 0 to 1. Then, by Laplace's rule of succession (8.6), the probability of k black balls is $1/61$ for $k = 0, \ldots, 60$, the same as before. The expected value of the number of black balls is 30, and if we repeat the auxiliary experiment, the marginal probability of drawing a black ball is $1/2$. For the whole urn, including the 30 red balls, the marginal probability of drawing a black ball is $1/2 \times 2/3 = 1/3$, the same as the probability of drawing a red ball.

Thus, if we treat the ambiguous probability as objective probability, then bets I and II are equivalent, as are bets III and IV. However, in a specific bet, the number of black balls could be 15 or 45, or any number between 0 and 60. While the number of red balls is predetermined at 30, the number of black balls is not determined, but only has an expectation of 30. That is, the number of black balls has another layer of uncertainty. However, a Bayesian who believes in his uniform prior would say that I and II are equivalent, so he is ambiguity neutral and ignores the extra uncertainty in a specific situation.

This paradox also highlights the descriptive and normative aspects of an axiomatic system. If the system is only descriptive, then we simply say that it does not predict well in this case. But if it's normative like classical logic or arithmetic, then the violators must reconsider and correct their choices.

Ellsberg (1961) reported various reactions among some prominent economists and statisticians. Some did not violate the principle (G. Debreu, R. Schlaifer, P. Samuelson). Some violated the principle 'cheerfully' (J. Marschak, N. Dalkey); others 'sadly and persistently, having looked into their hearts, found conflicts with the axioms and decided to satisfy their preferences and let the axioms satisfy themselves. Still others (H. Raiffa) tended, intuitively, to violate the axioms but felt guilty about it... .' Some who previously felt 'first-order commitment' to the principle were 'surprised and dismayed to find that they wished, in these situations, to violate the Sure-Thing Principle.' This special group 'seems to deserve respectful consideration' because it included none other than L. J. Savage himself.

Raiffa (1961) later tried to explain away the ambiguity aversion, calling it an error. He wrote:

> If most people behaved in a manner roughly consistent with Savage's theory then the theory would gain stature as a descriptive theory but would lose a good deal of its normative importance. We do not have to teach people what comes naturally.

It's a curious dogmatic stance, since we can't say that classical logic would ever lose its normative importance even if everyone follows it consistently. However, he described some interesting studies to justify his claim that ambiguity aversion is an error. We will come back to this below.

Explanation in terms of simultaneous bets

As with Allais's paradox, there is a setup where ambiguity aversion is clearly irrational. It requires one to play *two simultaneous bets* I vs II, followed by III vs IV based on a single draw from the urn. The preferences I>II and IV>III imply the composite bet I+IV > II+III. This is irrational since both I + IV and II + III are sure to win $100 regardless of which ball is drawn. This means that asking for two simultaneous bets changes the act qualitatively. Explicitly asking a rational and intelligent person to make two simultaneous bets based on a single draw will remove the ambiguity aversion.

TABLE 14.6
Ellsberg's paradox explained in terms of two simultaneous bets on the outcome of a single draw from the urn. Preferring I>II and IV>III implies (I+IV)>(II+III), which is irrational because they have exactly the same payoff (a sure gift of $100).

Option	Red	Black	Yellow
I+IV	100	100	100
II+III	100	100	100

Now we can explain Ellsberg's paradox as follows: The sure-thing principle implicitly presumes that the preferred acts are acted on together by one decision maker based on the single realization of the random outcome (or a single state of the world). It's clear that if each of the two pairs of bets I vs II and III vs IV are given to two independent persons, it does not make sense to expect them to coordinate their preferences according to the sure-thing principle. Each person may have their own subjective preferences.

What happens empirically seems to be that, when confronted with Ellsberg's setup and questions, a real person considers these two pairs of bets as independent single decisions. In such a situation, it's no longer obvious that the ambiguity aversion is connected to the sure-thing principle, although as we have described above, a Bayesian can be expected to be neutral towards ambiguity.

In response to the paradox, Raiffa (1961) described a modified version that's essentially a randomized version of the simultaneous bets I+IV and II+III. First, toss a coin once fairly. For the I+IV bet, choose bet I if the toss yields heads and bet IV if tails. For the II+III bet, choose bet II if the toss yields heads and bet III if tails.

In effect, only a single draw of the ball is used for the bet, and regardless of the relative number of black and yellow balls, the two composite bets I+IV and II+III give exactly the same payoff. This is because no matter which ball is drawn, you'll have a 50-50 chance of getting 0 or $100. Therefore, it does not make sense to prefer I+IV over II+III. However, it's clear that the randomization step has changed Ellsberg's original setup by tying the two bets together.

Randomized Ellsberg's paradox

Raiffa motivated the randomized step with a simpler version of the paradox that is meant to show that ambiguity aversion is a logical mistake. Suppose the urn contains only 60 balls of two colours, red and black. In Scenario I, it's known there are 30 balls of each colour; in Scenario II, the number of each colour is unknown. In each case, you first choose a colour, and then pick a ball from the urn. If your pre-chosen colour is correct, you get $100, otherwise you get nothing. When tried on the students, they preferred I over II (i.e., they're willing to pay more to bet on I than on II.) This reflects the same ambiguity aversion as in Ellsberg's setup.

A randomised step is then introduced in II. In effect, you let a fair toss of a coin decide for you: if heads, you choose red, otherwise black. Then, regardless of the colour composition of the balls in the urn, the probability of choosing the correct colour is 0.5. For instance, suppose the balls are all red; if you knew that (which you don't) you should of course pre-choose red and you'd win; by letting the coin decide you only have an *objective probability* 0.5 of winning. This argument is true for any number of black balls; if you find this surprising, see the short proof in this chapter's appendix. In effect, the randomization step just turns Scenario II into Scenario I, so it does not address the ambiguity at all.

In any case, can we use the randomization step as a separate principle to deal with ambiguity and complete ignorance? That is, can we use it as an independent logical support for the sure-thing principle? Again, by Sinn's (1980) theorem in Section 9.1, the ambiguous probability must be treated like an objective probability *if* we accept the sure-thing principle (and the weak ordering axiom). So, Raiffa's argument is circular: His randomization step depends on the sure-thing principle.

Appendix: Randomized Ellsberg's Paradox

Suppose the proportion of red balls in the urn is p. You let a fair coin toss decide for you: If heads, you choose red; otherwise, black. To win, you must either get heads and draw a red (R) ball, or get tails and draw a black (B) ball. Assuming that the toss and the ball draw are independent, the probability of winning is

$$0.5\Pr(R) + 0.5\Pr(B) = 0.5p + 0.5(1 - p) = 0.5,$$

regardless of p.

15

Fallacious Fallacies

We use probability to quantify uncertainties in our reasoning for making judgements and decisions. Intriguingly, when used mathematically, probability-based thinking can sometimes produce starkly different conclusions compared to when it is used statistically. Our aim in this chapter is to discuss a well-known fallacy called the conjunction fallacy and a related inclusion fallacy. It will highlight that lack of statistical reasoning may lead to unnecessary confusion in our discourse on human rationality and decision making.

The title of the chapter is taken from Hintikka (2004), who questioned whether the conjunction fallacy is a fallacy at all. In our daily communications, we routinely make conjunctional statements. In the spirit of this book, it might be considered a paradox, where seemingly valid probability-based reasoning leads to a contradiction viz-á-viz statistical reasoning.

After the pioneering works by Daniel Kahneman and Amos Tversky, for which Kahneman received the Nobel Memorial Prize in economic sciences in 2002, a whole field of behavioural economics has now grown that describes how humans do not always behave according to what is rational or optimal as prescribed by the rules of probability. Extensive examples are found in Kahneman's book *Thinking, Fast and Slow* (2011, TFS) or Richard Thaler's *Misbehaving* (2016). Are we then not rational?

We will discuss two different modes of reasoning, one captured by probability and the other by likelihood, and suggest that our seemingly irrational behaviour is often due to decision making based on the likelihood. Thus, from the likelihood perspective, we are still acting rationally. Recognizing these two modes may lead to a better understanding and assessment of our decisions. We will also describe the results of a fascinating study that showed traces of likelihood-based reasoning in 15-month-old infants, indicating that such a mode of reasoning is perhaps natural in human thinking.

15.1 Conjunction Fallacy

Also known as the Linda Problem, this fallacy originated from Tversky and Kahneman (1983):

> Linda is 31 years old, single, outspoken, and very bright. She majored in philosophy. As a student, she was deeply concerned with issues of discrimination and social justice, and also participated in anti-nuclear demonstrations. Which is more probable: T: Linda is a bank teller, or $T\&F$: Linda is a bank teller and is active in the feminist movement?

What is your answer? If you choose $T\&F$, then you're like most people. However, selecting $T\&F$ over T is supposedly contrary to logic. By the elementary conjunction rule of probability, $\Pr(T\&F) \leq \Pr(T)$, and, given any Data,

$$\Pr(T\&F|\text{Data}) \leq \Pr(T|\text{Data}),$$

so T is the safer choice. The phenomenon has been described in negative terms as a fallacy, sin of representativeness, cognitive bias, conjunctive violation, misunderstanding [of probabilities], etc.

From Kahneman (TFS, p. 158): 'About 85% to 90% of undergraduates at several major universities chose the second option, contrary to logic. Remarkably, the sinners seemed to have no shame. When I asked my large undergraduate class in some indignation, "Do you realize that you have violated an elementary logical rule?" someone in the back row shouted, "So what?"' After seeing the results of their empirical studies, he continued, 'I quickly called Amos [Tversky] in great excitement to tell him what we had found: "we had pitted logic against representativeness, and representativeness had won!"'

The evolutionary biologist S. J. Gould (1988) wrote 'I am particularly fond of this example because I know that the [second] statement is least probable, yet a little homunculus in my head continues to jump up and down, shouting at me: "but she can't just be a bank teller; read the description!".' As we shall see later, Gould should have listened to his little homunculus.

If you search the internet for 'conjunction fallacy,' you will even find courses that cover this fallacy as a warning of irrational human reasoning. So why do the majority of undergraduates – and the homunculus in Gould's head – willingly or even wilfully violate an 'elementary logical rule'? Many studies conducted by Hertwig and Gigerenzer (1999) pointed to the role of misunderstanding the question as a (partial) explanation of the failure. Misunderstanding the question can, for example, be due to misunderstanding the words 'probably' or the conjunctive 'and.' However, the term 'misunderstanding' presumes that 'probable' should only be interpreted mathematically in terms of probability.

In reality, it appears that most people interpret 'probably' in a broader sense, including in terms of statistical support and preference. Thus, the question about Linda can be posed as follows: Which statement is *more supported by the data*: T or $T\&F$? Technically, 'more supported' by the data may depend on higher likelihood or other confirmatory measures. The latter is discussed in Section 15.5 below. Alternatively, we might ask: Given the evidence, which statement *would you prefer*: T or $T\&F$? Higher preference depends not only on probability, but also on utility or rewards.

So, we shall focus on explanations based on statistical thinking and likelihood. Unfortunately, the Linda literature often does not make a distinction between 'probable' and 'likely', or between 'probability' and 'likelihood'. In their seminal 1983 paper, Tversky and Kahnemann themselves used 'probable' 39 times and 'likely' 24 times, each time having the same meaning. Some articles even use the term 'inverse probability' to refer to the likelihood. Not to be pedantic, we will also not make any distinction between 'probable' and 'likely,' but will always use 'likelihood' in its technical sense.

15.2 Inclusion Fallacy

A closely related phenomenon, called the 'inclusion fallacy' by Shafir et al. (1990), occurs where a general proposition is judged more likely than a more specific one. For instance:

H_1: All bank tellers are conservative.
H_2: All feminist bank tellers are conservative.

First of all, both statements are unlikely to be true. However, if H_1 is true, then H_2 must be true, so $\Pr(H_1) \leq \Pr(H_2)$, considering these as logical or subjective probabilities. Yet, most study subjects would rank H_1 as more likely than H_2. A different type of inclusion fallacy is the following: Given the evidence that robins use serotonin as a neurotransmitter, which is more likely:

H_1: All birds use serotonin as neurotransmitter.
H_2: Ostriches use serotonin as neurotransmitter.

Now, for a majority of study subjects, H_1 is judged more likely than H_2, even though H_1 implies H_2. So, as in the conjunction fallacy, does this indicate that intuitive human reasoning does not follow the mathematical probability rule? If so, what kind of reasoning does it follow?

15.3 Statistical and Scientific Reasoning

Before going into a more formal explanation, we first show that the so-called fallacies are, in fact, something that occurs regularly in statistical data analyses and scientific reasoning. They cease to be fallacies as soon as the propositions are treated as statistical hypotheses, where uncertainties due to incomplete information are acknowledged. Instead of being a fault of common reasoning, the alleged fallacies are perhaps a reassuring fact that normal humans follow sensible statistical and scientific thinking.

Let's start with a typical statistical data analysis involving hypothesis testing. We want to compare the mean blood glucose levels of 3 groups of people based on their weight: underweight, normal weight and overweight. Call the true means μ_1, μ_2 and μ_3, respectively. The following are the hypotheses along with their respective normal-based z-statistics and 2-sided P-values. While any detailed data are not provided, this information will still allow us to draw relevant conclusions:

$$
\begin{aligned}
H_1 &: \mu_1 = \mu_2 \qquad (z = 1.51, P = 0.13) \\
H_2 &: \mu_2 = \mu_3 \qquad (z = 1.18, P = 0.24) \\
H_3 &: \mu_1 = \mu_3 \qquad (z = 2.68, P = 0.0073).
\end{aligned}
$$

We would typically conclude that we cannot reject H_1 and H_2, but we can reject H_3 (at 1% significance). We could also say that H_1 and H_2 are more likely than H_3. But *mathematically*, H_1 and H_2 together imply H_3. So, we have the inclusion fallacy! When faced with the data, no scientist would ever take the mathematical implication seriously, because they know that their conclusions about H_1 and H_2 are uncertain. Therefore, instead of a fallacy, the apparent contradiction reflects our concession to uncertainty.

For the original inclusion-fallacy example in the previous section, we can write the hypotheses formally as

$$\begin{aligned} H_1: &\quad \theta_1 = 1 \\ H_2: &\quad \theta_2 = 1, \end{aligned}$$

where θ_1 is the proportion of bank tellers who are conservative, and θ_2 the proportion of feminist bank tellers who are conservative. Mathematically, H_1 implies H_2. Suppose a sample of bank tellers consists of only conservatives, so we can't reject H_1. But we would still suspect that, to qualify as good feminists, a sample of feminist bank tellers would reveal at least a few progressives. That would lead to the rejection of H_2. So, we arrive at the alleged inclusion fallacy, that H_1 is more likely than H_2, but again, no statistician would consider this a fallacy.

Now let's consider an informal, but statistically close and more realistic, statement of the hypotheses. If we're uncertain whether 'All bank tellers are conservative,' we could state it as '*Most* bank tellers are conservative.' But now the statement no longer implies that 'All feminist bank tellers are conservative.' It might even be the case, without any contradiction, that 'All feminist bank tellers are progressive.' This means that, by acknowledging the statistical uncertainty, H_1 no longer implies H_2, and they can be judged independently.

What's happening in the Linda Problem? The fallacy occurs as we are again snookered by a similar mathematical implication: Linda being a feminist bank teller implies she is a bank teller. However, as we are not certain that Linda is a bank teller or a feminist bank teller, we could express the uncertainty in the following statements:

T^*: Linda is typical of bank tellers.
$T\&F^*$: Linda is typical of feminist bank tellers.

These statements acknowledge explicitly that we are only guessing; who knows, Linda could be a politician or a journalist. Most importantly, now $T\&F^*$ no longer implies T^*, so we can judge that $T\&F^*$ is more likely than T^* without any contradiction.

15.4 Probability and Likelihood-Based Reasoning

The supposedly correct probability-based reasoning is to choose T over $T\&F$ regardless of the data. However, by completely ignoring the evidence, this reasoning actually looks dubious. If anything, the fallacy indicates that probability is not the only metric for comparing hypotheses. First, let's simplify the problem so that we can use exact numbers and put the Linda Problem in a theoretical framework where correct choices are more obvious.

Somebody picks a card at random from a standard deck of cards and you have to choose which is more likely:

> P: It's a picture card.
> $P\&S$: It's a picture of spades.

Both hypotheses have small probabilities, but here we're only interested in the relative comparison. A priori, without any clue on the card, you'd choose P as it is 4 times more likely than $P\&S$. Clearly, there is no preference for a conjunction if there is no information.

At the other extreme, suppose you get the clue ('Clue') that the card is a picture of spades. There shouldn't be any argument that the reasonable action is to choose $P\&S$ over P. This is, of course, a highly contrived situation, but it does indicate that a conjunction *can* be a rational choice given a sufficient amount of information. What's happening here in terms of the conditional probability and likelihood? They are:

$$\begin{aligned}
\Pr(P|\text{Clue}) &= 1, \\
\Pr(P\&S|\text{Clue}) &= 1, \\
L(P) &= \Pr(\text{Clue}|P) = 3/12, \\
L(P\&S) &= \Pr(\text{Clue}|P\&S) = 1.
\end{aligned}$$

Therefore, given the information, P and $P\&S$ are now equally probable: The prior gap between the probabilities of P and $P\&S$ has been removed by the data. The correct technical statement is to say neither is more probable than the other. Yet, it doesn't make sense to feel ambivalent between P and $P\&S$. There is a natural choice: $P\&S$, the one with the higher likelihood.

The more interesting and relevant question is what happens when you only have partial information. Suppose you're told that it's a picture card ('Picture'). Then, of course, you'd choose P over $P\&S$, as there is no information about the spade. In this case we have

$$\begin{aligned} \Pr(P|\text{Picture}) &= 1, \\ \Pr(P\&S|\text{Picture}) &= 3/12, \\ L(P) &= \Pr(\text{Picture}|P) = 1, \\ L(P\&S) &= \Pr(\text{Picture}|P\&S) = 1. \end{aligned}$$

Thus, the likelihood is neutral, but the probability clearly points to P as the better option.

Now consider getting this piece of information:

Black = the card is black.

Do you feel the temptation to choose $P\&S$ over P? Perhaps not. Intuitively, the information is not convincing enough. The conditional probabilities are

$$\begin{aligned} \Pr(P|\text{Black}) &= 6/26, \\ \Pr(P\&S|\text{Black}) &= 3/26, \end{aligned}$$

while the corresponding likelihoods are

$$\begin{aligned} L(P) &= \Pr(\text{Black}|P) = 1/2, \\ L(P\&S) &= \Pr(\text{Black}|P\&S) = 1. \end{aligned}$$

Here $P\&S$ has a higher likelihood than P, but P still has a higher probability than $P\&S$, so there is no obvious reason to prefer $P\&S$.

Finally, the most interesting case, the one closest to the Linda Problem: What if you get the information:

Spade = the card is a spade.

Intuitively, if you're going to consider a picture card at all, $P\&S$ becomes tempting as we know that the card is a spade. If you wilfully choose P, you'll be ignoring the spade information (sounds familiar?). The conditional probabilities and likelihoods are

$$\begin{aligned}\Pr(P|\text{Spade}) &= 3/13,\\ \Pr(P\&S|\text{Spade}) &= 3/13,\\ L(P) &= \Pr(\text{Spade}|P) = 3/12,\\ L(P\&S) &= \Pr(\text{Spade}|P\&S) = 1.\end{aligned}$$

The conditional probabilities are in fact the same, so rationally you can't use them for ranking: The information has closed the prior gap between the probabilities of P and $P\&S$. The temptation to choose $P\&S$ is consistent with its higher likelihood value as a tiebreaker.

Comparing Black and Spade, both are neutral regarding the hypothesis P:

$$\Pr(P) = \Pr(P|\text{Black}) = \Pr(P|\text{Spade}) = 3/13,$$

The key difference between them manifests itself in the new constituent hypothesis S:

$$\Pr(S|\text{Spade}) = 1 > \Pr(S|\text{Black}) = 1/2.$$

So, to make a conjunction a tempting proposition, the information must be highly convincing for the new constituent of the conjunction.

Returning to the Linda Problem, we can see a close parallel to the Spade case. Her description does not actually tell us that she is a bank teller, but reveals a person with progressive views. Let's see what happens if you consider the information convincing enough to predict that she is a feminist, i.e., $\Pr(F|\text{Data}) \approx 1$. (Isn't that the intention of the description?) Theoretically, this implies

$$\Pr(F|T\&\text{Data}) \approx 1,$$

meaning that with the added consideration that Linda is a bank teller, you would still predict that she is a feminist. Applying the multiplication rule,

$$\begin{aligned}\Pr(T\&F|\text{Data}) &= \Pr(T|\text{Data}) \times \Pr(F|T\&\text{Data})\\ &\approx \Pr(T|\text{Data}).\end{aligned} \tag{15.1}$$

Hence, as in the Spade case, the information has closed the prior gap between the probabilities of T and $T\&F$. We can no longer say that T is more probable than $T\&F$. So, the choice between T and $T\&F$ will be driven by the likelihoods

$$\begin{aligned} L(T) &\equiv \Pr(\text{Data}|T), \\ L(T\&F) &\equiv \Pr(\text{Data}|T\&F). \end{aligned}$$

Assuming $\Pr(T) > \Pr(T\&F)$, the posterior equality (15.1) implies

$$L(T\&F) > L(T),$$

so you will choose $T\&F$ over T. In summary, to those who believe, based on her description, that Linda is a feminist, the conjunction is a legitimate choice.

Similar reasoning applies to the medical judgment and Wimbledon examples from Tversky and Kahnemann (1983). Doctors know well that pulmonary embolism (blood clot in the lung) often leads to dyspnoea (breathing difficulty). Therefore, those who believe that dyspnoea is inevitable would consider the combination of dyspnoea and another symptom is more likely than that other symptom alone.

In the Wimbledon example, for those who strongly believed that Bjorn Borg would win the title in 1981, winning after losing the first set is still more likely than losing the first set. This example has been said to be different from the standard Linda Problem as it involves future events, and conditioning on future events to compute the likelihood is unnatural or even invalid. That's a misleading view, since none of the probability or likelihood calculations assume a causal relationship, so time direction is irrelevant.

However, the time element could give another potential explanation based on conditional probability. First, as the background data, Borg was ranked no.1 coming to Wimbledon in 1981 and had won it 5 times in a row, so it's reasonable to predict he'd win it again, i.e. $\Pr(W|\text{Data}) \approx 1$. As before, defining L_1 to mean losing the first set, we get

$$\Pr(L_1\&W|\text{Data}) \approx \Pr(L_1|\text{Data}),$$

again seemingly giving a neutral choice, but there is another potential tie-breaker. We know $\Pr(W|\text{Data}) \approx 1$ also implies $\Pr(W|L_1\&\text{Data}) \approx 1$, which is winning the title *after* losing the first set – only a very subtle difference from 'winning the title *and* losing the first set'. Whereas, no one would predict confidently that Borg would lose the first set, i.e. $\Pr(L_1|\text{Data}) < 1$. This may have led some people to choose $L_1\&W$ over L_1.

15.5 Degree of Confirmation

The reasoning used to arrive at the conjunction fallacy is a fallacy only if the words 'probable' or 'likely' are interpreted in a technical way as mathematical probability. This presumes that the probability itself is the only carrier of information in the data. There are, however, other metrics. Crupi et al. (2008) suggested that, in judging or comparing propositions, people intuitively use a degree of confirmation (Carnap, 1950) rather than the probability itself. For instance, our reasoning compares the *change* in probability due to evidence rather than the probability itself. Instead of comparing

$$\Pr(T|\text{Data}) \text{ vs } \Pr(T\&F|\text{Data})$$

we first compare the posterior probability with the prior probability, thus

$$\{\Pr(T|\text{Data}) - \Pr(T)\} \text{ vs } \{\Pr(T\&F|\text{Data}) - \Pr(T\&F)\}.$$

Or, alternatively, we compare the ratios

$$\frac{\Pr(T|\text{Data})}{\Pr(T)} \quad \text{vs} \quad \frac{\Pr(T\&F|\text{Data})}{\Pr(T\&F)}.$$

In the Linda Problem, the data are negatively correlated with T and positively correlated with F, so the left side is less than 1, while the right side is greater than 1. So, it's possible that people intuitively discount the likeliness of T, while inflating it for $T\&F$. Note, however, that in view of Bayes's theorem (8.2), these ratios will always have the same ordering as the likelihoods.

Other measures of confirmation have been proposed in the literature. Crupi et al. showed that, in general, all known measures of confirmation of a conjunction will be discordant with conditional probability – hence potentially leading to the fallacy – if evidence is negatively correlated with T but positively correlated with F. Furthermore, if the evidence is negatively correlated with T, then the measure of confirmation of $T\&F$ is concordant with the likelihood. However, the Spade example in the previous section shows that a negative correlation with T is not necessary, but we must have convincing evidence for F.

15.6 Traces of Likelihood-Based Reasoning in Infants

Gweon et al. (2010) performed a fascinating series of experiments on inductive learning by infants – average age 15 months old – as evidence that perhaps likelihood-based reasoning is hard-wired in our brains. The story is engagingly told in Laura Schulz's Ted Talk, found at `https://www.ted.com`.

How does a baby's logical mind work? Here, we only highlight their key experiment: The babies were presented with *3 squeaky blue balls* taken from a large opaque box. The box wall facing the babies had a clearly visible picture indicating its content. Two scenarios were tried according to what was shown in the picture:

> Scenario 1: Mostly blue balls and some yellow balls;
> Scenario 2: Mostly yellow balls with some blue balls.

The ratio is 3:1 in each case, but as far as the babies were concerned, we presume the exact number did not really matter. The blue balls were taken one at a time and, each time, the babies were shown that they squeaked. Then the babies were given a single yellow ball; the question was, would they attempt to make it squeak? What reasoning or inference method do they use?

First, we can agree that the only 'Data' available to the babies are {3 squeaky blue balls}. To use the probability-based reasoning, the babies would have to come up with the posterior probability

$$\text{Pr(Yellow balls are squeaky}|\text{Data}),$$

which would of course require the prior probability before seeing the data and some form of Bayesian calculation to compute the posterior.

This might be too difficult for the poor babies. The likelihood-based reasoning would require the babies to assess

$$L(\text{Yellow balls are squeaky}) = \Pr(\text{Data}|\text{yellow balls are squeaky}).$$

This is perhaps not obvious either, as there was never any direct evidence of the squeakiness of yellow balls; so, the inference must come from an inductive generalization. Gweon et al. (2010) described a model involving four hypotheses leading to predictions based on likelihood reasoning, but here we shall construct a simpler thought process. On seeing three squeaky blue balls, the babies were implicitly assessing these two hypotheses:

H_1: The sample is randomly selected from all the balls.
H_2: The sample is not random, but was selectively taken only from squeaky blue balls.

Furthermore, with category-based induction (see Section 2.5), when a sample is judged random then the properties of the sample will generalize more easily; for example, squeakiness then applies to all balls, hence to the yellow balls. And vice versa, when a sample was judged not random, then the properties would not generalize easily.

On observing the Data {3 squeaky blue balls}, the likelihoods of the hypotheses are now computable. For Scenario 1 (mostly blue balls):

$$\begin{aligned} L(H_1) &= \Pr(\text{Data}|H_1) \sim \text{ high} \\ L(H_2) &= \Pr(\text{Data}|H_2) \sim \text{ high.} \end{aligned}$$

The exact values are $0.75^3 (=0.42)$ and 1, respectively; presumably the babies did not use the exact values, but used only visual clues to conclude that there was no reason to reject the random sampling hypothesis H_1. Therefore, the squeaky property was generalized to yellow balls and the babies were predicted to squeak the yellow balls. On the other hand, for Scenario 2 (mostly yellow balls):

$$\begin{aligned} L(H_1) &= \Pr(\text{Data}|H_1) \sim \text{ low} \\ L(H_2) &= \Pr(\text{Data}|H_2) \sim \text{ high.} \end{aligned}$$

Again, the exact values are $0.25^3 (= 0.016)$ and 1, respectively. Using only visual clues to judge the likelihoods, the babies would reject the random sampling hypothesis H_1 and hence not generalize the squeaky property.

So they were predicted not to squeeze the yellow balls. The experiment confirmed these predictions, or we could also say that likelihood-based reasoning explains the experimental results, providing evidence of the elementary use of likelihood-based reasoning in infants.

15.7 Discussion: The Rewards of Being Right

In the first card example in Section 15.4, given the evidence, why does it feel natural to prefer $P\&S$ over P even if they have the same probability? In his comments, Kahneman (TFS, p.158) referred to the study participants violating an 'elementary logical rule.' Indeed, if you violate the normative rules of classical logic or arithmetic, you cannot escape the verdict of being illogical. But the reasoning under uncertainty is different: There are no universally accepted normative rules. The closest we have to normative rules is the axiomatic system developed by Savage in Chapter 7, where being rational implies the existence of probability and utility, *not probability alone.*

The utilities in relation to a hypothesis are how much we gain for being right and how much we pay for being wrong. In other words, the cost-benefit aspect. The probability of being right or wrong is not enough. In real betting, events with a lower probability will have higher prizes, so conjunction events could be more attractive than each constituent event. In science, scientists formulate and test the strongest hypothesis *supported by the data*, not the most probable one. The strongest hypothesis will in fact be the least probable a priori. Think about Einstein's theories of relativity, which contain the least expected propositions about space and time.

Thus, a hypothesis is valued for its explanatory and predictive power. The most probable hypothesis might be the blandest with the weakest explanatory power. 'A feminist bank teller' is a bank teller for sure, but the feminist element makes it a more interesting hypothesis with more explanatory and predictive power than the bland 'bank teller.' We could easily consider other hypotheses such as

W: Linda is a woman
H_u: Linda is human,

which would have much higher probabilities than T. On the other hand, in the Linda Problem, it is of course not reasonable to choose '$T\&F\&W_d$: Linda is a widowed feminist bank teller,' as there is nothing in the data telling us about the widowed status.

Another lesson we learn from the utility-based rationality theory is the crucial difference between single vs repeated events, because they're associated with different utilities. One's reasoning in a single bet is different from the reasoning in multiple bets; the long-term expected value may not be meaningful for a single event. If I ask you the question, which is larger: the number of bank tellers or the number of feminist bank tellers? It would indeed be illogical to say that the latter is larger than the former, regardless of what information you have. But when an individual named Linda comes into the picture, the question of which group she belongs to can legitimately be interpreted differently from the question about comparative group size.

We have given some theoretical explanations that provide a rationale of preferring the conjunction $T\&F$ over T, which depends in part on the likelihood. We are not claiming to know why most people choose to do that. It could be that the psychological rewards of being right for a good reason, i.e., by using available evidence, are attractive. But what we can say is that there are valid statistical or scientific reasons and that probability alone is not sufficient for reasoning under uncertainty.

The conjunction fallacy highlights the use of likelihood in reasoning. In general, why and when should one rely on likelihood? First, when prior probabilities are clearly unavailable or inappropriate. The Linda Problem shows another situation where the conditional probability appears to be uninformative. The infant study also illustrates how one can use likelihood for reasoning. In fact, we can use these examples to indicate that many people – including babies – appear to use such direct likelihood inference.

What do we gain from distinguishing probability-based reasoning from likelihood-based reasoning? Primarily, clarifying the meaning of the terminologies we use will also clarify our thinking, thereby reducing unnecessary confusions. For example, we believe there is no need to call the conjunction fallacy a fallacy and to accuse ourselves of being illogical or irrational, when we are actually using a legitimate likelihood-based reasoning. Closing the gap between technical meaning and layman understanding is always a difficult challenge in the public dissemination of science, but perhaps not hopeless. At the very least, the distinction between probability-based vs likelihood-based reasoning should be part of a standard scientific discourse on decision making.

16

Monty Hall Puzzle and the Three Prisoners Paradox

The Monty Hall puzzle has been solved and dissected in many ways, but always using probabilistic arguments, so it is considered a probability puzzle. Here the puzzle is set up as an orthodox statistical problem involving an unknown parameter, a probability model and an observation. This means we can compute a likelihood function, and the decision to switch corresponds to choosing the maximum likelihood solution. One advantage of the likelihood-based solution is that the reasoning applies to a single game, unaffected by the future plan of the host. We also describe an earlier version of the puzzle in terms of three prisoners: two to be executed, and one released. Unlike goats and cars, these prisoners are sentient beings that can think about exchanging punishments. When two of them do that, however, we have a paradox, where it is advantageous for both to exchange their punishment with each other. Overall, the puzzle and the paradox are useful examples of statistical thinking and modelling using likelihood.

16.1 The Puzzle and the Paradox

First, here is the Monty Hall puzzle:

> You are a contestant in a game show and are presented with three closed doors. Behind one is a car, and behind the others only goats. You pick one door (let's call that Door 1), and then the host *will* open another door that reveals a goat. With two unopened doors left, you are offered a switch. Should you switch from your initial choice?

It is well known that you should switch; doing so will increase your probability of winning the car from 1/3 to 2/3. The problem is actually how to convince those who think that there is no reason to switch because, for them, the chance of winning for either of the unopened doors is 50-50. The puzzle, first published by Selvin (1975ab), was inspired by a television game show *Let's Make a Deal* originally hosted by Monty Hall. It later gained widespread attention after appearing in Marilyn vos Savant's *Ask Marilyn* column in *Parade* magazine back in 1990. Her assertion that the optimal strategy was to switch doors led to numerous letters criticizing her advice. Many of the letters were signed by individuals with Ph.D. credentials, although some later conceded that vos Savant was correct. The puzzle has generated its own literature and many passionate arguments *even* among those who agree that you should switch.

Martin Gardner's 1959 article on the same puzzle in terms of three prisoners (below) was titled *Problems Involving Questions of Probability and Ambiguity.* He started by referring to the American polymath Charles Peirce's observation that in 'no other branch of mathematics is it so easy for experts to blunder as in probability theory.' The main problem is that some probability problems may contain subtle ambiguities. What seems so clear to us – as in the statement of the Monty Hall problem above – may contain hidden assumptions we forget to mention or are not even aware of. For example, as will become clear later, the standard solution assumes that if you happen to pick the winning door, then the host will open one of the other two doors randomly with probability 0.5. For some, that is obvious, in order to make the problem solvable and the solution neat. If we explicitly drop this assumption, then there is no neat solution, but our discussion is not going in that direction.

Instead, we head to a more interesting paradox. In Martin Gardner's version, there are three prisoners A, B and C, where two will be executed and one released. Being released must feel even better than winning a car. Suppose A asks the guard: Since for sure either B or C will be executed, there is no harm in telling me which one. The guard says that C will be executed. Should A ask to switch his punishment with B's? With the same logic as in the Monty Hall problem, it must be yes, A should switch with B. Now, B asks the same question to the guard: Which of A or C will be executed? The guard again answers C. Then it seems advantageous for B to switch with A. Now we have a paradox: How can it be advantageous for both A and B to switch?

We can make another version of the paradox: Suppose A just *hears the breaking news* that C will be executed (no guard is involved). Should A ask to switch his punishment with B? What logic should apply here? If it is the same as before, then he should switch. But then, the same logic applies to B, and we arrive at the same paradox. If the same logic does not apply, then having the guard answer the questions matters. But why does it matter how A finds out that C is to be executed?

In probability-based reasoning, you are comparing the probability of winning if you stay with your initial choice vs if you switch. This is relatively easy with the Monty Hall puzzle – though hard enough for some people, but it gets really challenging if we want to explain the three prisoners paradox. So, instead, the puzzle and the paradox will be set up as an orthodox statistical problem involving an unknown parameter, a probability model and an observation. This means we can compute a likelihood function, and the decision to switch corresponds to choosing the maximum likelihood solution. In the Monty Hall puzzle, one advantage of the likelihood-based solution is that the reasoning applies to a single game, unaffected by the future plan of the host. In addition, the likelihood construction will have to show explicitly all the technical assumptions.

16.2 Likelihood-based Solution

Let θ be the location of the car, which is completely unknown to you. Your choice is called Door 1; for you, the car could be anywhere, but, of course, the host knows where it is. Let y be the door opened by the host. So, the 'data' in this game is the choice of the host. Since he must avoid opening the prized door, y is affected by both θ and your choice. The probabilities of y under each θ are given in this table:

	$\theta = 1$	$\theta = 2$	$\theta = 3$	$\widehat{\theta}$
$y = 1$	0	0	0	–
2	0.5	0	1	3
3	0.5	1	0	2
Total	1	1	1	

For example, it is clear that y cannot be 1, because the host cannot open your door, so it has zero probability under any θ. If Door 1 is the winning door ($\theta = 1$), the host is *assumed* to choose randomly between Doors 2 and 3. If $\theta = 2$, he can only open Door 3. Finally, if $\theta = 3$, he can only open Door 2. Reading the table row-wise gives the likelihood function of θ for each y, so the maximum likelihood estimate (MLE) $\widehat{\theta}$ is obvious:

If $y = 2$, then $\widehat{\theta} = 3$ (so you should switch from 1 to 3).
If $y = 3$, then $\widehat{\theta} = 2$ (switch from 1 to 2).

The increase in the likelihood of winning by switching is actually only twofold, so it is not enormous.

In this likelihood formulation, the problem is a classic statistical problem with data following a model indexed by an unknown parameter. For orthodox non-Bayesians, there is no need to assume that the car is randomly located, so no probability is involved on θ; it is enough to assume that you know completely nothing about it. From the table, it is also clear that, when $\theta = 1$, the host does not need to randomize with 50-50 probability; he can use any other split, and you will never decrease your likelihood of winning by switching and sometimes increase it.

16.3 Probability or Likelihood?

Previous solutions to the Monty Hall puzzle are usually given in terms of probability. Now we're going to limit our discussion to frequentist probability. For non-Bayesians, why bother with likelihood? Imagine a forgetful host in the Monty Hall game: He forgets which door has the car, so he opens a door at random. If it reveals a goat, the game can go on; if it reveals a car, then he makes excuses, the game is cancelled and nobody wins. Suppose that in your particular game, a goat is revealed. Is it still better to switch?

As before, let's call your chosen door Door 1. Define the data y as the *unopened door* (other than yours) if the opened one is a goat (so the game is on); if the opened one is a car, then set $y \equiv 4$ (and the game is off). The probability table is now:

	$\theta = 1$	$\theta = 2$	$\theta = 3$	$\widehat{\theta}$
$y = 2$	0.5	0.5	0	{1,2}
3	0.5	0	0.5	{1,3}
4	0	0.5	0.5	–
Total	1	1	1	

So, when $y = 2$ or 3, and the game is still on, your likelihood of winning the car with your original or the other unopened door is equal, and there is no benefit of switching! Compared to the original version, this means that *the evidence of an open door revealing a goat is not sufficient to say that switching is beneficial.* We must also know whether the host's choice is intentional or accidental.

But what about the future plan of the host? Should it also affect your reasoning? In frequentist probability reasoning, it should also matter because probability is not meant to apply to a specific game. Say that in the current game the host *forgets the winning door and accidentally* opens a door with a goat, but for future games he *plans* to never forget and will always intentionally open the goat door. What logic applies to the current game? Non-Bayesians certainly cannot apply the orthodox probability-based argument to the current game. But likelihood-based reasoning still applies to the current game, unaffected by the future plan of the host.

16.4 Three Prisoners Paradox: A Guard Involved

The likelihood-based reasoning also helps solve the three prisoners paradox. Let θ be the identity of the prisoner to be released; to avoid confusion, set $\theta = a$, b or c for the three prisoners A, B and C. Let y_1 be the guard's answer to prisoner A, and denote the answer by a, b or c according to which prisoner he says will be executed. The guard cannot answer a to prisoner A, must never reveal who will be released and must randomize his answer whenever possible. Then, as shown below, we indeed have a similar probability table as for the game show. So, we get the same MLE as before, making switching a good strategy.

	$\theta = a$	$\theta = b$	$\theta = c$	$\widehat{\theta}$
$y_1 = a$	0	0	0	–
b	0.5	0	1	c
c	0.5	1	0	b
Total	1	1	1	

When B asks the same question, let y_2 be the guard's answer. Assume for now that A does not know about y_2, and similarly B about y_1. The joint probability distribution of (y_1, y_2) is as follows. It can be derived under the same requirements: The guard cannot tell the questioner that he would be executed, must never reveal who would be released and must randomize whenever possible.

	$y_1 = a$	$y_1 = b$	$y_1 = c$
$\theta = a, y_2 = a$	0	0	0
$y_2 = b$	0	0	0
$y_2 = c$	0	0.5	0.5
$\theta = b, y_2 = a$	0	0	0.5
$y_2 = b$	0	0	0
$y_2 = c$	0	0	0.5
$\theta = c, y_2 = a$	0	1	0
$y_2 = b$	0	0	0
$y_2 = c$	0	0	0

Keeping only the non-trivial scenarios, which contain at least one non-zero probability, the table can be simplified to

(y_1, y_2)	$\theta = a$	$\theta = b$	$\theta = c$	$\widehat{\theta}$	Better for A	Better for B
(b, a)	0	0	1	c	Switch with C	Switch with C
(b, c)	0.5	0	0	a	No switch	Switch with A
(c, a)	0	0.5	0	b	Switch with B	No switch
(c, c)	0.5	0.5	0	$\{a, b\}$	None	None
Total	1	1	1			

Based on the relevant outcome of the story ($y_1 = c, y_2 = c$), the prisoners A and B have an equal likelihood of being released. So, a switch confers no advantage on either side, and there is no paradox.

How does the paradox arise? Let's consider who has access to what information. A only knows $y_1 = c$, but does not know the existence of y_2, so his reasoning is incomplete. Similarly, B only knows y_2. Overall, the paradox arises due to incomplete information. An external agent who knows both answers y_1 and y_2 – e.g., the guard – can see that there is no advantage in switching.

However, if the guard's answers are to be shared as common knowledge between the guard, A and B, then from the last table we could see that ($y_1 = c, y_2 = c$) is the only acceptable outcome; all the other answers reveal who would be released. For ($y_1 = c, y_2 = c$), there is no advantage for A and B to switch. The paradox will occur if A and B ignore the common knowledge and each uses his own information only. (This is an interesting parallel to the situation in which two individuals decide to use distinct subjective probabilities when there exists an objective probability. Individually, A and B are coherent, but an external agent can take advantage of their ignorance of the common knowledge. See page 98.)

Finally, suppose that each of A and B knows that the other has asked the relevant question, but A only knows his own answer y_1 and B only knows y_2. Furthermore, assume that a switch can only happen by mutual consent. Is the information enough to stop them from wanting to switch with each other? Firstly, A would not want to switch if $y_1 = b$. If $y_1 = c$, then, for A, a switch is advantageous only when $y_2 = a$, but neutral when $y_2 = c$. But $y_2 = a$ means that B is told that A will be executed, so he will not volunteer to switch with A. The same logic applies to B. So, when both sides are willing to switch, they know that the switch must be neutral, and there is no paradox.

16.5 Three Prisoners Paradox: Breaking News

As before, let θ be the identity of the prisoner to be released. Now, assuming that the executions are in random order, let y be the identity of the first prisoner reported to be executed. Then we can derive the probabilities under different θ's.

	$\theta = a$	$\theta = b$	$\theta = c$	$\widehat{\theta}$
$y = a$	0	0.5	0.5	$\{b, c\}$
b	0.5	0	0.5	$\{a, c\}$
c	0.5	0.5	0	$\{a, b\}$
Total	1	1	1	

So, when $y = c$, A and B have equal likelihoods of being released. In other words, when both A and B hear that C will be executed, there is no advantage for either of them to switch punishment. We can also see a similarity with the previous discussion that if C's execution is a common knowledge between A and B, then there is no advantage in switching.

Why does the explanation look so trivial? Well, we can always make it more complicated. The fact that A hears the news means that he is not the first to be executed. Suppose that A was, in fact, told that if he is to be executed (which he might not), then he will not be the first. Does that affect how he should react to the news? Using the same definition for y as the first prisoner to be executed, the probability table becomes:

	$\theta = a$	$\theta = b$	$\theta = c$	$\widehat{\theta}$
$y = a$	0	0	0	–
b	0.5	0	1	c
c	0.5	1	0	b
Total	1	1	1	

Hey, this looks familiar! Of course, it is exactly the same table as for the original Monty Hall puzzle above. So, yes, in this case, the news is informative: A should indeed ask to switch with B. But what did they tell B? Suppose B knows that A is told that A will not be the first to be executed, but B himself is not told that. How should he react to the news? From the table, when $y = c$, then $\widehat{\theta} = b$, so B should *not* switch with A.

If both are told that they won't be the first to be executed – and both know this, then the table becomes:

	$\theta = a$	$\theta = b$	$\theta = c$	$\widehat{\theta}$
$y = a$	0	0	0	–
b	0	0	0	–
c	1	1	0	$\{a, b\}$

So, on hearing the breaking news that C is executed ($y = c$), A and B are back to the situation of neutral switch. (When both A and B are told that they are not the first to be executed, then C must be the first to be executed, but that cannot happen under $\theta = c$. That's why the probabilities are all zero for $\theta = c$.)

16.6 The Case of a Talkative Guard

At this point you could be feeling confident that you have seen all possibilities. But consider the following version: A asks the guard and is told that C will be executed. So, A would, of course, like to switch with B. However, the guard then says, 'now I'm going to B and tell him that C will be executed, but you (A) can still decide whether or not to switch with B regardless of B's consent.' This feels equivalent to the news version with the guard as the news reader, and A would realize that a switch makes no difference. Which reaction is correct? Is A's optimal decision dependent on knowing the future plan of the guard? Note that it could be just a plan, not necessarily acted on. Is simply being told that B will know that C will be executed enough for A to change his mind?

From our previous analysis in Section 16.4, to make the story reasonable – e.g., not revealing who would be released – there are many restrictions on what the guard could say to prisoners A and B. Even more constraints would appear if the guard wants to say more things to A.

Let y_1 be the guard's answer to A and y_2 the information that the guard tells A he would tell B. Suppose that, as we have previously assumed in the original version, if $\theta = a$ then the guard would randomize his answer y_1 between b and c.

The summary table on page 246 lists four probable values of (y_1, y_2) in the original setting. Clearly, $y_2 = a$ is no longer possible, as it would be telling A that he'd be executed. And if $y_1 = b$, the guard cannot say $y_2 = c$ either, as it would reveal to A that he would be released. So, if $y_1 = b$, his only choice is to make no additional comment, i.e., $y_2 = \cdot$. This leaves $(y_1 = b, y_2 = \cdot)$ and $(y_1 = c, y_2 = c)$ as the only permitted outcome. We can summarize the probabilities in the following table.

(y_1, y_2)	$\theta = a$	$\theta = b$	$\theta = c$	$\widehat{\theta}$
$(b, \cdot)$	0.5	0	1	c
(c, c)	0.5	1	0	b
Total	1	1	1	

Back to the familiar table again! Yes, it's like the one for the original Monty Hall puzzle and the first table on page 246. The entries under $\theta = b$ and $\theta = c$ are as before, since the guard cannot randomize. Hence, on hearing $(y_1 = c, y_2 = c)$, it's better for A to switch with B. The extra data $y_2 = c$ carries no information, since it's not optional given $y_1 = c$.

How does the confusing idea about the similarity with the breaking news version arise? It happens if the guard *does not* intend to randomize his answer and plans to tell A $(y_1 = c, y_2 = c)$ when $\theta = a$. So, the table becomes as given below. That's why the guard's answer feels like the breaking news and it's no longer advantageous for A to switch.

(y_1, y_2)	$\theta = a$	$\theta = b$	$\theta = c$	$\widehat{\theta}$
$(b, \cdot)$	0	0	1	c
(c, c)	1	1	0	$\{a, b\}$
Total	1	1	1	

In summary, both reactions by A can be correct under different assumptions on randomization. Confusion would arise if the guard's intention is not explicitly stated or its crucial role is not realized. This is similar to the effect of the host's intention in the original Monty Hall puzzle.

16.7 Discussion

Not surprisingly, optimal decisions by the player in the Monty Hall game or by the prisoners depend on the model setup and assumptions and the available data. Certain facts, e.g., an open Door 3 revealing a goat, are not sufficient on their own, as they can be interpreted differently depending on how we end up with them. People who think that switching makes no difference only consider the final stage of two closed doors. As in most statistical data analysis problems, how we collect the data matters. One notable lesson from the three prisoners paradox is how delicately the solution depends on the phrasing and fine details of the story.

One advantage of the likelihood-based approach is that it forces us to make all assumptions explicit, so it clarifies them. Another advantage compared to the standard frequentist probability-based reasoning is its applicability to a single game, unaffected by future intentions or plans of the host.

17

The Lottery Paradox and the Cold Suspect Puzzle

Dealing with uncertainty is never easy. Although probability as the main tool for dealing with uncertainty is well developed, we still have to deal with many probability-related puzzles and paradoxes. In this chapter, we describe a rather idiosyncratic selection that highlights the problem of accepting uncertain statements. Without going into a formal decision theory, there are simple intuitive rational bases for doing that, based, for instance, on high probability alone. The lottery paradox shows the logical problem of accepting uncertain statements based on high probability. The cold suspect puzzle arises in the quantification of forensic evidence in the assessment of suspects from database searches. The frequentists claim that the evidence is weaker than in the hot suspect case, the Bayesians say that it is stronger, while the likelihood analysis shows that neither side is right.

17.1 The Lottery Paradox

Since we all face uncertainty all the time, one would think that we should be good at dealing with it. But we are not. Why is that? First, consider Kyburg's (1961) rule of acceptance: It is rational to accept an uncertain proposition that has a high probability of being correct. This sounds reasonable, and one might argue that we cannot live without this rule. We go about our daily lives and make plans without thinking about the various risks that we know are constantly around us, as we cross the road, drive to work, take a plane journey, etc. We know accidents do happen, but with small enough probabilities that we assume that they will not happen to us. In fact, your family and friends will be concerned if you start worrying obsessively about minor risks in life.

Kyburg came up with the lottery paradox to highlight the logical problem of accepting uncertain propositions, even those with very high probability. There is a large body of literature surrounding the paradox; see the Wikipedia entry to get an overview. Suppose that you buy a ticket in a lottery where your chance of winning is 1 in a million. It is rational to go on living the same way as if you had not bought the ticket; you can't start planning a fancy car or a foreign holiday, etc., on the off chance of winning the lottery. In other words, you behave like you're going to lose. You would also expect every other lottery buyer to think the same way. Yet, invariably, there is a lottery winner. So, in effect, you're holding a contradiction: Each buyer is going to lose, but one – or even more than one – buyer is going to win.

A formal analysis of Kyburg's paradox is simple if we presume the lottery is a raffle, where each ticket number is sold at most once and there is a guarantee of one winner. Assuming, there are 100 tickets sold, the probability that a particular ticket i loses is

$$\Pr(\text{Ticket } i \text{ will lose}) = 99/100,$$

which is large enough to justify the feeling that you're going to lose. But, for a collection of k tickets, the probability that all of them will lose is

$$\Pr(k \text{ tickets will lose}) = (100 - k)/100, \ \text{ for } k = 1, \ldots, 100,$$

which drops to zero as k increases. At $k = 100$, the probability that all tickets will lose is zero, or a winner is guaranteed.

What makes acceptance of uncertain statements sometimes difficult or non-intuitive? Certain/sure statements are governed by the more intuitive but rigid and truth-preserving laws of deductive logic. For instance, if statements S_1 and S_2 are individually true, then the joint statement $S_1 \& S_2$ must be true. We can combine any number of statements with *no degradation of certainty*. That is not the case with uncertain statements: A conjunction will generally degrade the overall quality.

In the lottery example, each statement that the ticket i will lose has a probability of 0.99, but the probability that two tickets will jointly lose is degraded to 98/100. The probability continues to degrade with larger and larger joint statements until it goes to zero altogether. Our willingness to hold the contradictory positions – each will lose but somebody will win – is our concession to uncertainty. It's something we would be embarrassed to do when dealing with certainty.

That willingness might also be an indication that our brains are not wired to think probabilistically, as it is easy to get lost in the conjunctive tangles of a joint event. In the so-called birthday paradox, first imagine two people comparing their birthdays. The probability that they share a birthday is 1/365. Therefore, it seems reasonable to presume that they will not celebrate their birthdays together. When a third person comes, the probability of any pair of them sharing a birthday is also 1/365, and so on for more and more people. But, considering all the pairs together, the chance of some people sharing birthdays will increase, but by how much? For instance, what happens in a room of, say, 25 people? Is it likely that two or more people will share a birthday? It turns out to be quite likely; the probability has increased to 0.57. In a room of 50 people, it is almost certain (probability 0.97) that some people can celebrate their birthdays together. Most people will find this unintuitive.

17.2 Probability in Court

Uncertainty rules in courtrooms, but at some point judges or juries have to accept and assimilate the uncertain statements provided by the prosecution and the defence, and make appropriate decisions. What acceptance rules do they follow? The basis of conviction in criminal cases is 'beyond reasonable doubt.' What does that mean? It should be the case that the evidence provided is sufficient to remove any lingering doubt. Does that mean a level of probability of guilt so high that any remaining doubt is unreasonable? In civil cases, the standard of proof is 'the balance of probability.' This sounds like a probability greater than 0.5, but in any case, it definitely sounds less demanding than 'beyond reasonable doubt.' As shown in the case of O.J. Simpson's murder trial, an acquittal in a criminal court might be followed by a loss in a civil court.

The use of probability in court raises inherently difficult issues directly connected to the meaning of probability. First, let's consider probability as a long-term frequency: How does it apply to the case at hand? For example, suppose it is known (statistically) in a nameless corner of the world that 60% of self-employed businessmen cheat on their taxes. Mr. John is a self-employed businessman. On the balance of probability he is a tax cheater, but no judge will convict him on the basis of statistical probability alone. If the statistical probability is sufficient for conviction, then every self-employed businessman should be convicted immediately. In the frequentist interpretation, the probability 60% does not apply to Mr. John. His defence lawyer can easily produce all sorts of facts as evidence of his good character, e.g., he is a faithful churchgoer, among whom the proportion of tax dodgers is negligible. Other evidence would have to be provided by the state to make a convincing case, e.g., is he living beyond his means? He declared an average income of $15,000 per year for the past few years, yet he lives in a big house and recently bought several fancy cars, etc.

In Bayesian philosophy (Section 6.5), probability is a subjective measure that applies to individual cases, but it is based on a subjective bet. Subjective probability is the price you would be happy to pay to win $1.00 if your bet is correct. Strictly speaking, from a betting perspective, an agent can make a bet based on whatever information is available to him. In the tax-cheating example, suppose you're the judge. *If there is no other information available to you*, then the statistical probability of 0.60 is the fair price for a bet that Mr. John is a cheater. If you're happy with the price, then you will bet that he is guilty of tax evasion. We're pretty sure that judges and juries are not encouraged to think in this way.

The full decision process by a jury is perhaps too complex to be captured by a single number such as the probability the defendant is guilty. In *The Probable and the Provable*, Cohen (1971) gave examples to indicate why a high probability alone is not a sufficient reason to convict. A probability is something offered by an expert witness, typically for a piece of evidence, and not something estimated by a juror as a measure of the quality of his/her decision. Here is an example: Adam has been accused of murdering Bob, and one piece of evidence is a threatening letter found at Bob's house. It was not signed, but an expert witness identified three unusual characteristics in the typed letter that could be produced by Adam's typewriter.

Suppose that the chance that a randomly chosen typewriter could produce all three characteristics together is 1 in a million. Therefore, there is a high likelihood that the letter was typed on Adam's typewriter. But there are still many other questions that the jury will want to know or that the defence lawyers will raise to keep reasonable doubts in the mind of the jury. Did Adam actually type the letter? Could there be someone else who used the typewriter in trying to frame him? If he did, did he do it voluntarily? And, anyway, how did the letter establish him as a murderer? Did he have the opportunity to kill Bob? And if he had, did he take advantage of it? Perhaps he had a strong alibi? Thus, there is a wide gap between the high probability of a piece of circumstantial evidence and the removal of reasonable doubt. To some extent, the criminal court demands an explanatory theory comparable to that in the court of science.

17.3 DNA Fingerprinting

DNA profiling, first developed in the mid-1980s by the geneticist Alec Jeffreys, is now commonly used in courts. It is a forensic technique to establish the 'genetic fingerprint' of any biological sample – such as blood, semen or hair – collected from a person or a crime scene. DNA profiling has also been used in civil cases, such as parental disputes, but to be specific, we shall use the language of a criminal court. There are many technical issues in the statistical analysis of the data (e.g., Roeder, 1994), but to make a long story short, the final evidence is typically presented to the jury in terms of a likelihood ratio (Section 10.3).

The numerator likelihood is the probability of a match between the DNA profile of the crime-scene sample and the DNA profile of the suspect. It is typically assumed that the test is fully sensitive, which means that if the suspect is the true source of the sample, the test will produce a positive result. So, the likelihood is

$$L(\text{Suspect is the source}) = \Pr(\text{DNA match}|\text{Suspect is the source}) = 1.$$

The denominator likelihood is the probability of a match if the suspect is not the source. A perfect test should produce a negative result, i.e., the probability of a match is zero. This is now possible with whole-genome sequencing, but it is still not in common use. The most commonly used DNA profiling is based on a limited number of DNA markers; since 2017, forensic labs in the US have used a set of 20 markers.[1] Because of this, a person's DNA profile is not fully specific: There could be person(s) in the background population – where the suspect belongs – who have a DNA

[1] `https://www.nist.gov/news-events/news/2016/12/nist-research-enables-enhanced-dna-fingerprints`.

profile similar to the suspect. This depends on statistical information on the DNA profile in that relevant population. So, the probability of a match is the probability that a random person has a DNA profile that matches the DNA profile of the crime sample. Although not zero, this probability is typically really small; estimating the size is part of the statistical technicalities we refer to above. So, in general

$$\begin{aligned} L(\text{Suspect not the source}) &= \Pr(\text{DNA match}|\text{Suspect not the source}) \\ &= p, \end{aligned}$$

where p is usually a tiny number, such as 10^{-9} or even much lower. The DNA evidence is then presented as the likelihood ratio

$$R_1 = \frac{L(\text{Suspect is the source})}{L(\text{Suspect not the source})} = 1/p, \qquad (17.1)$$

so R_1 is an impressively big number, like a billion or more.

With a DNA match, the jury could then be presented with statements like 'the suspect is a billion times more likely to be guilty than not guilty.' But, as for the typewriter case, the likelihood is actually not the likelihood of guilt directly. The defence would still ask, and the jury would want to know, for example: Where is the evidence that the suspect was actually at the crime scene? Otherwise, someone – God forbid, the police – might have tried to frame him. Or, the suspect claimed that he was drinking with friends. If there was a weapon involved, where is it? Is there evidence that the suspect had the weapon before the murder? Therefore, the jury may have reasonable doubts, not because the presented probability did not reach certainty, but because there is more specific evidence that they are still missing.

17.4 Database Search: The Cold Suspect Puzzle

When DNA profiling is performed on a hot suspect, there is little controversy about the use of the likelihood ratio R_1 in (17.1) to represent the level of evidence against the suspect. But how do you assess the evidence if the police have no suspect, but instead trawl through a database containing the DNA profiles of many individuals? The US Federal Bureau of

Investigation maintains a database of DNA profiles called CODIS (Combined DNA Index System). By April 2021, it contained approximately 20 M profiles from (i) convicted felons (>14 M profiles), (ii) persons who had ever been arrested (>4 M profiles), (iii) forensic samples from crime scenes (>1 M profiles) and (iv) relatives of missing persons. Since its inception in 1998, CODIS has assisted in more than 500,000 investigations.[2]

To see that there is an issue in assessing evidence based on a database search, let's imagine a card trick: A magician shows a standard deck of playing cards and asks you to shuffle them. He puts the deck face down, chooses one card, and puts it aside, still face down, so no one can see what it is. He asks you to think of a card and then name it out loud. You say 'ten of diamonds!' The magician then reveals his pre-chosen card, and, sure enough, it is the ten of diamonds! Everyone is duly impressed; the probability is 1 in 52 for the magician to select the correct card. This is what happens when a hot suspect, who is like the pre-chosen card, matches the crime sample. The small probability of a match explains the magical element of surprise experienced by the spectators.

If the magician simply searches for the ten of diamonds among the 52 cards and shows it, no spectator will be impressed. This is what happens in a cold search. Having no suspect, the police would send the DNA profile from the crime scene to CODIS for a search. If there is a good match, then the Matcher will be considered a suspect – let's call him a cold suspect – for a closer investigation.

Here is the puzzle: Is the DNA evidence against the cold suspect from the DNA profile weaker than the evidence against a hot suspect? If we follow the intuition from the card trick, then the answer seems to be yes. In the card trick, the evidence is weakened by 52 folds, so there is no surprise at all. Suppose the search is limited to convicted felons, so there are roughly 14 M profiles. Is the evidence weakened by 14 M folds? This would negate much of the strength of the DNA evidence.

Consistent with the reasoning above, an expert committee of the US National Research Council (NRC, 1996) recommended that the likelihood ratio R_1 from a cold search be divided by the size of the database, so that the strength of the evidence is greatly reduced. We may call

[2]`https://www.fbi.gov/news/pressrel/press-releases/the-fbis-combined-dna-index-system-codis-hits-major-milestone` and `https://www.fbi.gov/services/laboratory/biometric-analysis/codis`

this the frequentist solution. However, statisticians on the Bayesian side claim that the cold search has no negative effect on the strength of DNA evidence. (See, for instance, the Discussion by Dawid in Lindley, 2000, and Donnelly and Friedman, 1999, which we'll refer to as DF99). In fact, they would even go further: The evidence is now *stronger* than in the hot-suspect case, because the database search has ruled out all other suspects. Who is right? The likelihood-based reasoning indicates that neither is.

The frequentist solution

Let $D+1$ be the number of profiles in the database and E be the event that there is a *single match* between the DNA profile of the crime sample and the profiles in the database. Let's call the person the Matcher. SID is the hypothesis that the source of the crime sample is in the database (you may think of the source as the criminal, but, as we have discussed, legally the guilty status is yet to be decided); SND is the hypothesis that the source is not in the database. As before, assume that the DNA profiling is fully sensitive, such that

$$\Pr(E|\text{SID}) = 1 \times (1-p)^D,$$

where p is the probability of a DNA match for a random non-source person. Given SND, the event E is a binomial event with probability

$$\Pr(E|\text{SND}) = (D+1)p(1-p)^D.$$

As a key note, using the binomial probability here means that we're only using the information that there is a single match in the database, even though we *actually know who the Matcher is.* The likelihood ratio is

$$R_F = \frac{(1-p)^D}{(D+1)p(1-p)^D} = \frac{R_1}{D+1} \leq R_1, \qquad (17.2)$$

where R_1 is the likelihood ratio (17.1) in the hot-suspect case. So, the evidence in the cold search is weaker than in the hot search, and that is in line with the NRC recommendation. It also corresponds to our intuition in the magic card trick: There is nothing magical if a match is found after searching the deck of cards. (DF99 presented another frequentist solution based on different probability reasoning that ends with the same likelihood ratio as an approximation.)

The Bayesian solution

The main criticism of the frequentist solution from the Bayesian side is that R_F represents evidence that is not relevant to the court, that 'somebody in the database is the source of the crime sample.' What matters to the court is whether a particular person – the Matcher – is the source. According to DF99 the likelihood ratio should be

$$R_B = \frac{R_1}{\Pr(\text{SND})} \geq R_1, \tag{17.3}$$

where Pr(SND) is the prior probability, before any DNA evidence, that the source not in the database; for convenience, the derivation is given in Appendix 1 of this chapter. This prior probability might depend, for instance, on the size, quality and relevance of the database. The likelihood ratio R_B is generally *larger* than the ratio in the hot pursuit, thus representing stronger evidence.

DF99 attributed this feature to the fact that many potential suspects have been ruled out. R_B also depends on the size of the database, but in a completely opposite way to the frequentist R_F. As the database gets larger R_F gets smaller, but Pr(SND) naturally drops, so R_B becomes larger. If the database is so large that it covers all potential suspects then $\Pr(\text{SND}) = 0$, i.e., the source is guaranteed to be in the database and the Matcher is guaranteed to be the source. This corresponds to $R_B = \infty$, which makes sense. This scenario has been highlighted by DF99 as a virtue of R_B and a weakness of R_F.

But what would happen at the other extreme, when $\Pr(\text{SND}) = 1$? That is, for whatever reasons, we have a priori knowledge that the source cannot be in the database; for instance, the database is a database of dead criminals or newborns. In view of (17.3) we have $R_B = R_1$, the ratio in the hot-pursuit case. But that does not make sense: In this case, we know that any Matcher must be a false positive, so the ratio should be zero.

The likelihood-based solution

Both the frequentist and Bayesian solutions have some weaknesses. We will now present a likelihood-based solution that overcomes them and highlights the evidential value of the likelihood. What bothers frequentists in the cold search is that, in contrast to the hot search, we do not specify any hypotheses in advance, i.e., before seeing the match. In terms of the card trick, the magician's pre-selection of the card makes the feat impressive, while trawling through the deck of cards after a card is mentioned looks decidedly unmagical.

However, we *can* specify the hypotheses *in advance.* We just have to specify many of them, namely one for each person. Let θ_i be an unknown binary status of whether person/profile i is the source of the crime sample, for $i = 1, \ldots, D+1$. The data y are the matching results. As before, assume that there is a *single match*, and let's call this matcher Person 1. The main hypotheses of interest are

(i) the Matcher is the source: $\theta_{11} \equiv (\theta_1 = 1, \theta_2 = 0, \ldots, \theta_{D+1} = 0)$,
(ii) the Matcher is not a source: $\theta_{10} \equiv (\theta_1 = 0, \ldots, \theta_{D+1} = 0)$.

In Appendix 2 of this chapter, we derive the likelihood ratio

$$R_L = L(\theta_{11}; y)/L(\theta_{10}; y) = 1/p = R_1 \tag{17.4}$$

for the Matcher being the source vs not being the source, exactly the same likelihood ratio as in the hot-pursuit case!

Assuming that we want a likelihood ratio that *purely represents the evidence in the data*, the DNA evidence in a cold search is the same as in the hot search. For some people, this result might seem intuitively obvious, but it is surprising to others. We will explain more below.

One surprising result in likelihood theory is that optional stopping has no impact on the likelihood (Section 10.5). So, one might imagine checking the DNA profile one at a time until a match is found (and assuming there is only a single match, the search can be abandoned once a match is found). This will produce the same likelihood ratio R_1 as above. The equivalence of the likelihood in hot and cold searches can be seen as the same phenomenon.

As in DF99, before the DNA search, we may have a prior likelihood ratio that the source of the crime sample is in the database vs not in the database. (Let's put aside the legal issue of whether the court would allow a formal prior assessment.) Using the same notation as DF99, the ratio can be written as

$$R_0 = \Pr(\text{SID})/\Pr(\text{SND}),$$

where $\Pr(\text{SID}) \equiv 1 - \Pr(\text{SND})$. Since we're not in the Bayesian framework, SND is a fixed hypothesis that does not have a probability and we should interpret Pr(SND) as prior likelihood, not prior probability. Now, if the source is in the database, then the Matcher is the source, and vice versa, if the source is not in the database, then the Matcher is not the source. So SID is equivalent to the hypothesis that the Matcher is the source, and SND equivalent to the Matcher not being the source. So the prior likelihood ratio is

$$L_0(\theta_{11})/L_0(\theta_{10}) = \text{Pr(SID)}/\text{Pr(SND)} = R_0,$$

and the total likelihood ratio including the prior likelihood is

$$R_T = R_0 \times R_1. \tag{17.5}$$

Here are some notable features of the total likelihood ratio: Unlike R_F or R_B, in general, R_T can be larger or smaller than R_1. The formula makes it explicit that the pure contribution of the data to the likelihood is in R_1, and that there is an independent contribution of prior information in R_0. When nothing is assumed about the database (or the court does not allow prior information), then we have the default $R_0 = 1$, so $R_T = R_1$. In contrast, since $R_0 = 1$ corresponds to $\text{Pr(SND)} = 0.5$, the Bayesian solution in this case gives $R_B = 2R_1$, which is rather non-intuitive.

As usual, the choice of the prior likelihood ratio can be controversial. One might consider, for example, the conviction rate of the people in the database compared to the rate in the general population. Extreme cases might also be argued: If the database is so large, e.g., including the whole population, that it is guaranteed to contain the source, then $\text{Pr(SND)} = 0$, so $R_0 = \infty$ and $R_T = \infty = R_B$. This makes sense as the Matcher is guaranteed to be the source. At the other extreme, when the database is irrelevant, so that $\text{Pr(SND)} = 1$, we have $R_0 = 0$ and $R_T = 0$, which is as expected, since evidence of a DNA match has no value. But in this case $R_B = R_1$, which does not make sense.

The card trick revisited

How do we interpret the fact that the likelihood ratio R_L ignores the evidentiary effect of the database search? In the card trick example, it is equivalent to equating the non-magician and magician routines. Let's consider another card example. You're shown a deck of cards facing down. *Before* you pick a card at random, you're told that the deck of cards is either

> H_0: a standard deck of 52 distinct cards, or
> H_{10}: a special deck containing 52 cards of tens of diamonds only.

Then you pick a card at random, and it's a ten of diamonds. Which kind of deck is more likely? You'd of course be suspicious that the deck has been rigged, so think that H_{10} is 52 times more likely than H_0.

Now, suppose you're not told *in advance* about the cards, but just pick a random card and see a ten of diamonds. Based on this information, which is more likely: H_{10} or H_0? Actually, exactly the same as before: H_{10} is 52 times more likely than H_0. This feels disturbing, since the hypothesis is specified *after* seeing the data. But that's not an issue. As for the likelihood ratio R_L above, we can specify all the possible decks in advance. So, the evidence in the data for the two scenarios is actually the same. What is different, and this affects our sense of surprise, is our prior expectation.

Our reaction and full processing uncertainty include both the prior expectation and the likelihood extracted from the data. The mismatch between the two generates a sense of surprise. It can be created by the magician's skill in handling the cards, which is then rewarded by the audience's applause. In the court case, the jury would likely feel the same way. Evidence of a DNA match against a hot suspect would feel like strong evidence that would impress the jury. But, like the magician, the police must have the competence to find pieces of evidence against the suspect, even just in order to identify him as a suspect. In contrast, a mere DNA match from a database will likely not convince the jury. It will only be a starting point for the police to build up the case, as they need to find other pieces of evidence in order to establish guilt beyond reasonable doubt.

Let's imagine two films about the same crime: one with the narrative going forward, starting with multiple clues that lead to the suspect, which is eventually supported by a DNA match. The other film goes backward, starting with a DNA match from a database search that identifies a potential suspect. The police then find multiple clues that establish the person as the suspect in the trial. Do you feel any difference in the strength of the total evidence against the suspect in the two narratives?

17.5 Discussion

It is, of course, not surprising that uncertain statements generate logical problems not found in certain statements. The lottery paradox highlights one such problem, where we knowingly hold contradictory positions: We rationally accept uncertain statements that have a high probability of being true, but we also happily accept the opposite of the consequences of those statements. The DNA forensic evidence story shows the limits of the use of probability-based reasoning in court. Putting aside the issue of the differential quantification of evidence by the different schools of inference, the resulting probability/likelihood assessment is not of the guilty status of the suspect. Probability is limited to an assessment of specific evidence. The full decision-making process of the jury, which must assimilate many more pieces of evidence, is too complex to be summarized into a numerical probability value.

Appendix 1: Derivation of R_B

For convenience, we show here the derivation of the likelihood ratio R_B (17.3) for the cold search in Donnelly and Friedman (1999). As before, let $(D+1)$ be the number of persons (=profiles) in the database; N be the number of potential suspects in the population outside the database; m be a factor such that a person in the database is m-times more probable to be the source relative to a person outside the database; and the evidence E be the recorded event that there is a single match in the database by a *particular person* called the Matcher.

The numerator of the likelihood ratio is the probability of E given that the Matcher is the source of the crime sample. This is the same as the probability in (17.5), which is $(1-p)^D$. The denominator is the probability of E given that the Matcher is not the source. This probability is computed as a weighted average

$$\frac{mD\mathrm{Pr}(E|S_d) + N\mathrm{Pr}(E|S_p)}{mD+N},$$

where '$\Pr(E|S_d)$ is the probability that the evidence would arise if someone other than the Matcher represented in the database were the source of the crime sample and $\Pr(E|S_p)$ is the probability that the evidence would arise if someone in the suspect population not represented in the database were the source.' The weights are the prior probabilities that the source is inside or outside the database, respectively, before the search.

The probability $\Pr(E|S_d) = 0$ as E is impossible under the assumption that the DNA test is fully sensitive. And, if the source is outside the database,

$$\Pr(E|S_p) = p \times (1-p)^D,$$

exactly the same meaning and value as the probability (17.5). So, the likelihood ratio is

$$\begin{aligned} R_B &= \frac{(1-p)^D}{Np \times (1-p)^D/(mD+N)} \\ &= \frac{R_1}{N/(mD+N)} = \frac{R_1}{\Pr(\mathrm{SND})}, \end{aligned}$$

which is the hot-suspect likelihood ratio R_1 divided by the prior probability that the source is outside the database before the search. As $N \to 0$ or $m \to \infty$, the database is guaranteed to contain the suspect, so $\Pr(\mathrm{SND}) \to 0$ and $R_B \to \infty$. When $m = 0$, the source must be outside the database, so $\Pr(\mathrm{SND}) = 1$, but $R_B = R_1$.

Appendix 2: Derivation of R_L

Let θ_i be an unknown binary status of whether person/profile i is the source of the crime sample, for $i = 1, \ldots, D+1$. Let $\theta = (\theta_1, \ldots, \theta_{D+1})$ the vector of the parameters. Define (i) a collection of $(D+1)$ hypotheses: Person i is the source for $i = 1, \ldots, D+1$, and (ii) the null hypothesis that the source is not in the database. Formally, in terms of the parameters, the $(D+1)$ hypotheses are $(\theta_i = 1, \theta_j = 0, j \neq i)$ for $i = 1, \ldots, D+1$, and the null hypothesis is $(\theta_1 = 0, \ldots, \theta_{D+1} = 0)$.

Let $y = (y_1, \ldots, y_{D+1})$ be the observed match between the profiles in the database and the crime-sample profile. As before, assume that there is a *single match*, and, without loss of generality, let's call this matcher Person 1. So, we have observed the data: $y_1 = 1$ and $y_i = 0$ for $i > 1$. The main hypothesis of interest is that Person 1 is the one and only source, $\theta_{11} \equiv (\theta_1 = 1, \theta_2 = 0, \ldots, \theta_{D+1} = 0)$. The likelihood is

$$L(\theta_{11}; y) = P_{\theta_{11}}(y_1 = 1, y_2 = 0, \ldots, y_{D+1} = 0) = 1 \times (1-p)^D.$$

Any competing hypotheses with $\theta_i = 1$ for $i \neq 1$ have zero likelihood by the assumption of full sensitivity of the test, so they are ruled out. The only other non-zero likelihood is for $\theta_{10} \equiv (\theta_1 = 0, \ldots, \theta_{D+1} = 0)$, so the Matcher is not a source. It is

$$L(\theta_{10}; y) = P_{\theta_{10}}(y_1 = 1, y_2 = 0, \ldots, y_{D+1} = 0) = p \times (1-p)^D.$$

So we end up with the likelihood ratio

$$\begin{aligned} R_L &\equiv L(\theta_{11}; y)/L(\theta_{10}; y) \\ &= 1/p = R_1 \end{aligned}$$

for the Matcher being the source vs not being the source.

18

Paradox of the Ravens

In Hempel's paradox of the ravens, using a seemingly logical reasoning, seeing a non-black non-raven – such as a red pencil – is considered evidence that all ravens are black. The paradox and its many resolutions indicate that we should not underestimate the logical and statistical thinking needed to assess evidence in support of a hypothesis. Most of the previous analyses of the paradox are within the Bayesian framework. These analyses, and Hempel himself, generally agree with the paradoxical conclusion. It feels paradoxical supposedly because the amount of evidence is extremely small. Here, we describe a non-Bayesian analysis of various statistical models with accompanying likelihood-based reasoning. The analysis shows that the paradox feels paradoxical because there are natural models in which observing a red pencil tells us nothing about the colour of ravens. In general, the value of the evidence depends crucially on the statistical sampling scheme and assumptions about the underlying parameters of the relevant model.

18.1 The Paradox

According to Nicod's criterion in philosophy, observing a black raven is evidence to support the statement that all ravens are black. We're not talking about the strength of the evidence here, but only whether it's supporting or not. The criterion further states that observing a non-black non-raven is irrelevant to the statement. This seems reasonable, but the philosopher Carl Hempel (1945b) came up with a well-known paradoxical counterexample. First, the following two statements are logically equivalent:

R1: All ravens are black.
R2: All non-black things are non-raven.

By Nicod's criterion, seeing a non-black non-raven, for instance a red pencil, is evidence for R2. Therefore, by equivalence, observing a red pencil is evidence that all ravens are black. This violates the criterion as well as our common sense. How can observing a red pencil tell us anything about the colour of ravens? As we walk around, we see an enormous number of non-black non-raven objects. Intuitively, they tell us nothing about the colour of ravens. Moreover, using the same logic, observing a red pencil also provides evidence for 'All ravens are white.' What is wrong with this reasoning?

The literature on the ravens paradox is surprisingly large. Surprising, because you would think that resolving such a small paradox should be easy enough. Yet, philosophers have argued about this for many decades. A peek at the Wikipedia entry would give you an overview of the many proposed resolutions and a more complete set of references. It is not our aim to review the existing literature. As a paradox of confirmation, the ravens paradox is of general interest because it raises a basic question about what it means to have supporting evidence. Hempel himself accepted the paradoxical conclusion and declared that the impression of a paradox is a 'psychological illusion.' According to him, the illusion arises because we have so much prior knowledge about a red pencil that its evidential support for the non-black non-raven property is somehow lost.

Many authors, perhaps guided by the Bayesian analyses that we will describe later, in fact accept the paradoxical conclusion. The paradox occurs supposedly because the confirmatory evidence is so tiny, as there are innumerable non-black objects compared with the obviously limited number of ravens. In any case, all the theories developed so far actually assume countable objects. To avoid the innumerability issue, let's consider old and young people, and their vaccination status. Let's call a person young if their age is less than 40, so that there is no ambiguity in what young means. The following two statements are logically equivalent:

S1: All old people are vaccinated.
S2: All unvaccinated people are young.

If observing an unvaccinated young person is supporting evidence for S2, does it also support S1? If you accept the reasoning of the ravens paradox, you must also accept that seeing an unvaccinated young person is evidence that all old people are vaccinated.

Most of the previous statistical analyses of the paradox are within the Bayesian framework. The only non-Bayesian analysis we know of is by Royall (1997). Our analysis here is an extension of Royall's by providing additional statistical models and the accompanying likelihood-based reasoning. The key result is that the paradox feels paradoxical because there are natural models where observing a red pencil has no relevance to the colour of ravens. Explicit technical-statistical assumptions, which are not so natural, are needed for the paradoxical conclusion to hold.

18.2 Data, Models and Their Likelihoods

To make our reasoning precise, we shall express it in a standard statistical framework. Viewed as a statistical problem, we wish to infer something about the vaccination status of old people based on observing an unvaccinated young person. We shall see that that is impossible without certain technical assumptions about the sampling scheme and the underlying parameters of the population.

In the simplest and perhaps most natural model, a random sample of size n is taken from a population of an unknown number N people. Two binary variables are recorded (y_1, y_2), with y_1 = age (old-young) and y_2 = vaccination status (yes-no). The unknown parameters are $N_1, \ldots, N_4$, presented in Table 18.1 ($\sum N_i = N$). Assuming there are old people, they are all vaccinated if and only if $N_3 = 0$. The corresponding observed data are $(n_1, \ldots, n_4)$ in the right table ($\sum n_i = n$). Seeing an unvaccinated young person can be represented as having a sample of size 1 and observing $(n_1 = n_2 = n_3 = 0, n_4 = 1)$.

TABLE 18.1
Simplest model: In the left table, $N_1, \ldots, N_4$ are the unknown population parameters representing the number of people in the categories defined by age and vaccination status. A random sample of size n is taken from the population. The right table shows the observed number of individuals $(n_1, \ldots, n_4)$ in the sample.

	Old	Young
Vax	N_1	N_2
NoVax	N_3	N_4

	Old	Young
Vax	n_1	n_2
NoVax	n_3	n_4

Hypergeometric model

Theoretically, according to the likelihood principle (Chapter 10), evidence in the data is captured by the likelihood function. On observing $(n_1, \ldots, n_4)$, the likelihood is based on the hypergeometric probability:

$$L(N_1, \ldots, N_4) = \frac{\binom{N_1}{n_1} \cdots \binom{N_4}{n_4}}{\binom{N}{n}}.$$

Specifically, for $(n_1 = n_2 = n_3 = 0, n_4 = 1)$, the likelihood is

$$L(N_1, \ldots, N_4) = N_4/N. \tag{18.1}$$

Clearly, the likelihood gives no information about the relative proportions of N_1 and N_3. The MLE is $(\widehat{N}_1 = \widehat{N}_2 = \widehat{N}_3 = 0, \widehat{N}_4 = N)$, at which point the likelihood is equal to 1 and zero everywhere else. So, observing an unvaccinated young person is *not evidence that all old people are vaccinated.*

Poisson model

The Poisson model offers a more transparent view of the problem. Let's assume that the number of people in the sample n_i is independent Poisson with mean $c\lambda_i, i = 1, \ldots, 4$, where λ_i is the unknown population number and c is the sampling ratio; see Table 18.2. The sample size is $n = c\sum_i \lambda_i$. Statements S1 and S2 can be expressed in terms of the parameters: S1: $\lambda_1 > \lambda_3 = 0$ and S2: $\lambda_4 > \lambda_3 = 0$.

TABLE 18.2
Two possible models and their parameterizations: (P) The number of individuals in each table entry is independent Poisson with means $\lambda_1, \ldots, \lambda_4$. (M) The observed number of sampled individuals in each cell of the table is multinomial with parameters n and $(\theta_1, \ldots, \theta_4)$, where $\sum_i \theta_i = 1$.

(P)	Old	Young
Vax	λ_1	λ_2
NoVax	λ_3	λ_4

(M)	Old	Young
Vax	θ_1	θ_2
NoVax	θ_3	θ_4

On observing $(n_1, \ldots, n_4)$ the likelihood is

$$L(\lambda_1, \ldots, \lambda_4) = \prod_{i=1}^{4} \Pr(Y_i = n_i),$$

where Y_i is Poisson with mean $c\lambda_i$. It is clear that the observed number of unvaccinated young people n_4 carries no information about the unknown number of vaccinated and unvaccinated old folks (λ_1 and λ_3). The MLEs are

$$\widehat{\lambda}_i = n_i/c.$$

If we have no idea of the total population size, then we do not know what c is, and the sample is not informative about the absolute size of λ_is; though it carries information about ratios of λ_is, captured in a multinomial model, which we describe next.

18.3 Multinomial Models for Various Sampling Plans

For large N_is the hypergeometric model is well approximated by the multinomial with sample size n and probabilities $(\theta_1, \ldots, \theta_4)$, where $\theta_i = N_i/N$; see Table 18.2. Similarly, conditional on the sample size, the Poisson model becomes a multinomial model with the parameters $(\theta_1, \ldots, \theta_4)$, where $\theta_i = \lambda_i / \sum \lambda_j$. Overall, the multinomial model is both convenient and transparent, and we shall use it from now on. In terms of θs, the statements S1 and S2 can be expressed thus: S1: $\theta_1 > \theta_3 = 0$ and S2: $\theta_4 > \theta_3 = 0$.

There is a crucial technical issue missing in many discussions about the paradox: How do we end up with the observation of an unvaccinated young person? Consider several possibilities, depending on how we take a sample:

Plan A: Take a random person from the total population list and ascertain their age and vaccination status.
Plan B: Take a random person from the list of young people and ascertain their vaccination status.
Plan C: Take a random sample from the list of unvaccinated people and ascertain their age.

Unlike in the ravens paradox setup, exact age and vaccination status are not visible qualities of the people, so logistically the plans are distinct from each other. When suggesting the paradox, Hempel, who was not a statistician, was obviously not thinking about sampling plans. But since these sampling plans have strong impact on the conclusion, let's consider them carefully. We discuss later which plan is most natural to model anecdotal observations such as black ravens or red pencils.

It is also perhaps not immediately obvious that assuming prior knowledge of the proportions $\theta_1, \ldots, \theta_4$ will have impact on our reasoning and conclusion. For example, suppose that the proportions of young people that are vaccinated and those that are not vaccinated are known (say $\theta_2 = 0.05$ and $\theta_4 = 0.45$, with exact values being unimportant), but the corresponding proportions of the old (θ_1 and θ_3) are not known. Then, intuitively, observing an unvaccinated young person will carry no information about the vaccination-status of the old. This is theoretically clear in the following subsections. So, explicit assumptions about what proportions are known are necessary for a definite solution.

Sampling plan A: completely unknown proportions

First consider the case where we assume ignorance of both the proportions of old people and vaccinated people. In other words, both margins in Table 18.2(M) are free. The parameters are only constrained by $(\theta_1+\theta_2+\theta_3+\theta_4) = 1$; so, we have three free parameters. On observing an unvaccinated young person ($n_1 = n_2 = n_3 = 0, n_4 = 1$), the likelihood is based on the multinomial probability:

$$L(\theta_1, \theta_2, \theta_3, \theta_4) = \theta_1^0\theta_2^0\theta_3^0\theta_4^1 = \theta_4.$$

Clearly, the likelihood contains no information about the relative proportions of vaccination in the old (the parameters θ_1 and θ_3). The MLE is $\widehat{\theta}_4 = 1$, which implies $\widehat{\theta}_1 = \widehat{\theta}_2 = \widehat{\theta}_3 = 0$. Hence, if seeing an arbitrary unvaccinated young person is taken formally as a random observation from the population where nothing is known about the age and vaccination status, *the observation is not evidence in support of S1.*

Sampling plan A: known age distribution

Now we start with the assumption that there is an equal proportion of old and young people, but we know nothing about their vaccination status; see Table 18.3. On observing ($n_1 = n_2 = n_3 = 0, n_4 = 1$), the likelihood is the same as before

$$L(\theta_1, \theta_2, \theta_3, \theta_4) = \theta_1^0\theta_2^0\theta_3^0\theta_4^1 = \theta_4,$$

in the parameter space constrained by $\theta_1 + \theta_3 = 0.5$ and $\theta_2 + \theta_4 = 0.5$; the space has two free parameters, say θ_3 and θ_4. Again, the likelihood is non-informative about the proportion of vaccinated and unvaccinated old people.

TABLE 18.3

The parameters θ_1 and θ_2 represent the proportions of old and young people, and similarly θ_3 and θ_4, for vaccinated and unvaccinated people respectively. By assumption, $\theta_1 + \theta_3 = 0.5$ and $\theta_2 + \theta_4 = 0.5$.

	Old	Young
Vax	θ_1	θ_2
NoVax	θ_3	θ_4
Total	0.5	0.5

Note that the argument *does not* depend on the assumption that old and young people have equal proportions; any proportions will do. Seeing an unvaccinated young person only tells us that there are some unvaccinated people ($\theta_3 + \theta_4 > 0$). In fact, the MLE here is $\widehat{\theta}_4 = 0.5$, which is of course not very precise, but the imprecision does not affect our argument. The constraint implies $\widehat{\theta}_2 = 0$, i.e., all young people are unvaccinated. Thus, seeing an unvaccinated young person is evidence that all young people are unvaccinated, but this is not equivalent to S2.

Sampling plan A: known age and vaccination-status distributions

Finally, suppose we know both the proportions of old people and of vaccinated people. Table 18.4 shows two possible sets of proportions. In Table 18.4(T1), the proportion of vaccinated people is 0.2 and that of unvaccinated people is 0.8. Since $\theta_3+\theta_4 = 0.8$, and each of θ_3 or θ_4 has a maximum of 0.5, both θ_3 and θ_4 must be at least 0.3. So, the statement S1 is trivially false.

To make the problem non-trivial, we must assume that the proportion of vaccinated people is at least 0.5; in general, this should be at least the

TABLE 18.4
Two possible sets of marginal proportions of vaccinated and unvaccinated people. Under sampling plan A, Sub-table T1 trivially falsifies S1, since θ_3 *must be at least* 0.3. *If the proportion of vaccinated people (0.2) is less than the proportion of old people (0.5), there must be unvaccinated old people in the population. In contrast, Sub-table T2 allows evidence supporting S1.*

(T1)	Old	Young	Total
Vax	θ_1	θ_2	0.2
NoVax	θ_3	θ_4	0.8
Total	0.5	0.5	1.0

(T2)	Old	Young	Total
Vax	θ_1	θ_2	0.6
NoVax	θ_3	θ_4	0.4
Total	0.5	0.5	1.0

proportion of old people. We will take Table 18.4(T2) as an example; different proportions that satisfy the non-triviality condition will lead to the same conclusion.

Given all the marginal constraints on $\theta_1, \ldots, \theta_4$, there is one free parameter left. For example, we can take θ_4 as the free parameter; in Table 18.4(T2), $0 < \theta_4 < 0.4$. On observing an unvaccinated young person, the likelihood is based on the multinomial probability as before:

$$L(\theta_1, \theta_2, \theta_3, \theta_4) = \theta_1^0 \theta_2^0 \theta_3^0 \theta_4^1 = \theta_4,$$

but under extra constraints $\theta_1 + \theta_2 = 0.6$ and $\theta_3 + \theta_4 = 0.4$. Because there is only one free parameter: As θ_4 varies from 0 to 0.4, the parameter θ_3 varies from 0.4 to 0. *Now the likelihood is informative* for each of the parameters. The MLE is $\widehat{\theta}_4 = 0.4$, which implies $\widehat{\theta}_1 = 0.5, \widehat{\theta}_2 = 0.1$ and $\widehat{\theta}_3 = 0$. In this scenario, observing an unvaccinated young person *does give evidential support for all old people being vaccinated.*

Incidentally, this supportive scenario corresponds to the standard Bayesian resolution (e.g., Good, 1960), where a prior distribution is assumed only on the number of non-black ravens (= unvaccinated old people), while *all the margins are fixed.* Perhaps not so obvious, this prior implies that there is only one free unknown parameter. Recall that, in the context of Bayesian inference (Chapter 8), 'free' means available for update by the data. The problem is that the supposed evidence in the non-black non-raven is then fully dependent on the choice of prior.

The Wikipedia entry on the paradox indicates that Bayesian analyses generally support the paradoxical conclusion. Good (1960) initially wrote – incorrectly – that the conclusion does not depend on the assumption of known marginals. He corrected it later (Good, 1961), but unfortunately, it seems the correction is not so well known. The Wikipedia entry referred to his 1960 paper in great detail, but made no mention of the 1961 paper.

Sampling plan A: summary

To summarize the results of plan A, which looks like a natural model for the anecdotal observation of an unvaccinated young person, the evidence generally does not support S1. Support for S1 only happens under the assumption that the proportion of vaccinated people and the proportion of old people are both known, and that the former is larger than the latter. This supportive scenario corresponds to the standard Bayesian resolution.

Sampling plan B

Under plan B, sampling young people and recording their vaccination status, we have a Bernoulli event with probability $p \equiv \theta_4/(\theta_2 + \theta_4)$ of seeing an unvaccinated one, so the likelihood is

$$L(p) = p.$$

It is clear from this that we can say something about the vaccination status of young people: that some, or perhaps all, of the young are unvaccinated. (In fact, the MLE $\widehat{p} = 1$). However, the likelihood carries no information about θ_1 or θ_3, so the evidence supports neither S2 nor S1. However, unlike plan A, the conclusion is not affected by additional information on the proportions of old people or vaccinated people, as given in Table 18.4.

Sampling plan C

Finally, under plan C, among the unvaccinated people, the proportion of young people is $p = \theta_4/(\theta_3+\theta_4)$. So $p = 1$ if and only if $\theta_3 = 0$. Observing an unvaccinated young person is a Bernoulli event with probability p, so the likelihood is

$$L(p) = p.$$

Now, the MLE $\widehat{p} = 1$ is consistent with $\widehat{\theta}_3 = 0$, meaning *the evidence does support S2 and S1.* As for plan B, this conclusion is not affected by the extra information about the proportions of old people or vaccinated people. We just need to assume that old people exist.

18.4 Large-Sample Evidence

Even when the observation is supportive, the evidence based on one unvaccinated young person of course feels unconvincing. However, in this thought experiment, the strength of the evidence is not the issue. What matters is whether the observation is supportive or not. Yet it is instructive to imagine a much larger sample size, such as 10,000, as the evidence should accumulate enough to feel more conclusive.

So, let's assume that, for each sampling plan: (i) we take a random sample of size not 1 but 10,000, and (ii) we only observe unvaccinated young people in the sample. Computationally, we just need to apply the power of 10,000 to the probabilities in the likelihood functions, so that we get highly concentrated likelihoods. However, we can perhaps follow the logic more intuitively using only verbal explanations. To be nontrivial, unless it's ruled out by the data, let's assume that there are old people in the population.

Under plan A, the evidence is consistent with the situation where the only people in the target population are unvaccinated young people. This means we cannot say anything meaningful about old people. It would be absurd to suppose that if there were old people, all of them would be vaccinated. If old people do exist in a non-trivial proportion, then the sample cannot have been taken randomly from the population, so again, we cannot infer anything about the vaccination status of the old. Overall, under plan A, observing even 10,000 unvaccinated young people does not tell us anything about the old people's vaccination status.

Under plan B, we sample 10,000 people from the list of young people and ascertain that all of them are unvaccinated. We have strong evidence for the statement 'All young people are unvaccinated', but that's not logically equivalent to S2. We cannot make statements about all unvaccinated people: Since we only sample young people, we do not know if there are unvaccinated old people. Consequently, the evidence supports neither S2 nor S1.

Under plan C, we sample 10,000 people from the list of unvaccinated people and ascertain that all of them happen to be young. This evidence strongly supports S2. It also tells us there are no old people among the unvaccinated, which can only be because all old people are vaccinated. *Hence, the evidence indeed supports S1.*

In summary, only under sampling plan C does the evidence of seeing 10,000 unvaccinated young people support the statement that all old people are vaccinated.

18.5 Discussion

Our analysis shows that Hempel's paradox seems paradoxical because there are natural scenarios where observing a red pencil – or even thousands of them – tells us nothing about the colour of ravens. Unlike Hempel's or the other previous analyses, the current analysis conforms to our common sense that there is no reason to accept the paradoxical conclusion. In an assessment of evidence, an explicit statement on how the observation was made and some technical-statistical assumptions about the underlying parameters are always needed. Perhaps obvious to scientists, supporting a statement or hypothesis requires a systematic gathering of evidence. Hempel's ravens paradox highlights the pitfalls in trying to interpret results from haphazardly collected data. It is interesting that our antennae are usually highly tuned for problems that arise with such data, but when put in a different context, as in Hempel's paradox, the problem somehow turns into a serious puzzle that generates lengthy debates.

We have gone through different probability models to capture the observation that underlies the paradox. The multinomial model under different sampling plans captures the problem most conveniently and transparently. Since the different sampling plans can lead to different conclusions, which is the most natural or appropriate? On first impression, sampling plan A, where we sample from the total population, seems to be the most appropriate. But under this plan, the paradoxical conclusion holds only if we assume that both of the marginal proportions are known. This assumption seems unlikely to hold as we casually move around and see myriad objects in our environment. Plan C, where we sample from unvaccinated people to determine their age, is the safest plan to establish supporting evidence for S1. But this plan requires a careful phrasing or reporting of the evidence, which does not conform to the original anecdotal phrasing of the paradox.

How come we do not have such sampling issues when we observe direct evidence of black ravens? It seems we could just walk around seeing more and more black ravens and becoming more and more convinced that all ravens are black. In our mind, we are naturally, but unconsciously, gathering data on ravens and keeping track of their colour. We can see this by our own reporting (if someone asks): We would say 'all the ravens we have seen so far are black'. Statistically, this does support the statement that all ravens are black. But nobody will ever report 'all the non-black things we saw were non-ravens,' because there is no meaningful data gathering on the properties of 'non-black things.' We could as easily be absorbed by the properties of non-ravens, which corresponds to plan B, hence giving no information as to the colour of ravens.

Nicod's criterion vs plausible reasoning

Hempel's paradox is a counterexample of Nicod's criterion. As the existence of a paradox indicates something is wrong, we see that the problem lies all along in the dubious scientific value of Nicod's criterion, because it ignores sampling plan. Would the paradox also occur in plausible reasoning (Section 8.6)? This is an important question because plausible reasoning is a cornerstone of scientific reasoning: We have a theory G that predicts a specific evidential instance E, i.e., a conditional statement 'G implies E.' Then, ascertaining E in an experiment increases the plausibility of G. Let G be the statement 'All old people are vaccinated,' and E be the statement 'we take a random old person and ascertain that he is vaccinated.' Clearly, G implies E, and in this case, E does increase the plausibility of G.

However, the statement 'We take a random vaccinated person and verify that he is old' is *not* implied/guaranteed by G, because there could also be vaccinated young people. Therefore, simply observing a vaccinated old person without any explanation of how they came under observation does not increase the plausibility of G. The correct application of plausible reasoning in this case also requires a proper sampling plan to provide evidence for G. In other words, Nicod's criterion does not follow the rule of plausible reasoning, and Hempel's paradox is not a counterexample of plausible reasoning.

19

The Exchange Paradox

We have explored the various interpretations of probability in Chapter 6. One intriguing paradox that arises in the context of subjective probability for single, non-repeated events is the exchange paradox, where an obviously neutral exchange is shown to be theoretically advantageous. It also highlights the role of likelihood, extended likelihood and confidence as measures of uncertainty for single events. By examining the paradox and its variants, we will gain more insight into the limitations of intuitive probability reasoning. Moreover, the expected utility theory will appear as an explanation of how we navigate uncertain situations. Overall, our discussion of the exchange paradox serves as a thought-provoking exploration of the interplay between probability, likelihood, utility theory and decision making.

19.1 The Paradox and Its Variants

Also known as the two-envelope paradox, the exchange paradox has an extensive literature. In the popular literature, Martin Gardner discussed it in his column on Mathematical Games in *Scientific American*, and in his 1982 book *Aha! Gotcha: Paradoxes to Puzzle and Delight.* Marilyn vos Savant wrote it in her 1992 *Ask Marilyn* column in *Parade* magazine and in her 1997 book *The Power of Logical Thinking.* In the statistical literature, the paradox has been discussed, for example, in Christensen and Utts (1992) from the Bayesian perspective and in Pawitan and Lee (2017) from the likelihood perspective. There is also a lengthy Wikipedia entry on the paradox, indicating a fascination among people of diverse backgrounds. Here is a typical setup.

Exchange paradox: An unknown θ dollar is placed in one envelope and 2θ dollars in another. You are asked to pick one envelope at random, open it and then are offered to exchange it with the other envelope. So, pick one at random and open it to reveal $y = \$100$. Now you reason: 'The content of the other envelope is either \$200 or \$50 with a 50-50 chance; If I exchange it, my expected winning is $(200+50)/2 = 125 > 100$. So, I should exchange the envelope, shouldn't I?' However, since this is true for any value of y, *there is no need to open the envelope, and you would still want to exchange it*, which is of course absurd since you have just chosen it randomly.

To avoid trivialities, it is worth emphasizing that we are not trying to 'resolve' the paradox by showing that, *in the long run*, exchanging cannot improve winning. This is trivial since, by random selection, the expected value is $(2\theta + \theta)/2 = 3\theta/2$ regardless of whether you exchange or not, and therefore there is no advantage in exchanging. (At least for now, since later we will modify the story to make exchanging an irresistible temptation.) We're also not going to discuss any psychological aspects by assuming, for instance, any trick or information on the limited wealth of the person who puts the money in the envelope, etc. We are focusing on the state of uncertainty in a single game and how it affects our decision making.

We next describe the 'all-or-nothing' (AON) version: Instead of choosing one of the envelopes, you are offered either both envelopes or none with equal chance. So, let's consider these three options to play the envelope game:

A. Keep: Choose one envelope at random and keep it.
B. Exchange: Choose one envelop randomly, then exchange it.
C. AON: Get either both envelopes or none with equal chance.

It is intuitively obvious that these options are mathematically equivalent, each having an expected value of $3\theta/2$. So, no option has any advantage over the others. But now we have the following:

All-or-nothing exchange paradox: Start as in the exchange paradox, where you pick one envelope at random and open it to reveal $y = \$100$. Then you're offered the following option: (i) keep your \$100, or (ii) take the AON option, where you compare it with the amount in the other envelope (still closed): If yours is smaller, you will get all the money, but if yours is bigger, you will lose your \$100. So, you reason: the content of the other envelope is either \$200 or \$50, so you will end up with either \$300 or nothing. The expected value is now $(300 + 0)/2 = 150 > 125 > 100$. This is much better than keeping the envelope and still better than exchanging, so you will of course take this AON offer, won't you? Since this is true for any value of y, *there is no need to open the envelope, and you would still want to take the AON option*, which is absurd since we just discussed that the AON option has no obvious advantage over the other options.

This AON option can be recognized as a version of the famous wallet game paradox (Section 12.4). Its appearance here only strengthens the warning that there is something suspicious with the intuitive probability-based reasoning. But we're only starting; there are more paradoxes to come.

In the simple exchange paradox, you're asked to pick one envelope and open it, and see $y = \$100$. We now change which envelope to open and end up with a different paradox:

Reverse exchange paradox: Start with the exchange paradox setup, but now you keep your envelope closed and instead open *the unchosen envelope*. Suppose that you see that the other envelope contains $y =$ \$100, so you reason: 'ok, then my envelope must have either \$200 or \$50 with 50-50 chance. So the expected value is \$125. Well, well, I better keep it!' Again, the reasoning applies to any value y, so you don't need to know the content of the other envelope to decide that your initial choice has a better prospect than the other envelope. That's, of course, absurd because you have just picked one envelope at random.

So, by simply changing which envelope to open, you can end up with a completely different – but equally paradoxical – recommendation. The AON option can also be modified in a similar way to get a reversed recommendation.

19.2 Before vs After Opening the Envelope

The exchange paradox is truly paradoxical *before* we open the envelope, as it is clearly illogical to want to exchange after randomly choosing one envelope and expect to get more money. The paradox must be due to treating our sense of uncertainty as probability. But aren't we axiomatically allowed to have subjective probability? Where is the specific point where the probability argument goes awry? We will answer this in the next section, but for now, the presence of the paradox suggests that you should perhaps not be tempted to exchange. But wait, life is not that simple.

Opening the envelope and changing the amount of money can make the temptation to exchange irresistible, as follows:

> **Magnified exchange paradox**: Instead of θ and 2θ, suppose that the envelopes contain θ and 100θ. You pick one randomly and open it to reveal $y = \$100$. So, you're thinking: 'the other envelope is either \$1 or \$10,000. By exchanging, I will basically either lose my \$100 or get \$10,000, with a 50-50 chance or whatever. I don't care what the reasoning is, but I want to exchange!'

The exact size 100 of the multiplier is not important; what matters is that it's big enough to tempt you to exchange. What is important is that your reasoning is affected by the amount you see. In the original paradox (with amounts θ and 2θ), if you happen to see $y = \$10,000$ and exchanging gives you the chance to win \$20,000, you may feel lucky enough and decide not to exchange, since you'd hate to lose that. The contestants in the game show 'Who Wants to Be a Millionaire' must know the feeling when they have to decide whether to continue the game or not after winning, say, \$10,000.

So, once you open the envelope, we don't really have a paradox anymore: The decision to exchange or not becomes a legitimate personal decision with no right or wrong judgement. The ultimate explanation of the exchange paradox must consider this magnified version.

19.3 Likelihood, Frequentist and Bayesian Explanations

Likelihood-based explanations

There are actually two distinct but closely connected likelihood-based explanations. One is based on the classical likelihood, and the other on the extended likelihood (Chapter 12). We start with the former and consider only the original exchange paradox; a similar argument applies for the AON and reverse exchange paradoxes.

If we knew absolutely nothing about θ, then on observing $y = 100$ we can only have $\theta = 50$ or 100 with likelihoods

$$\begin{aligned} L(\theta = 50) &= \Pr(Y = 100|\theta = 50) = 0.5 \\ L(\theta = 100) &= \Pr(Y = 100|\theta = 100) = 0.5. \end{aligned}$$

That is, the allowed values of θ have the same likelihood, not probability. If $\theta = 50$ then the amount in the other envelope must be 50. Otherwise, $\theta = 100$ and the other envelope is 200. Crucially, they have *equal likelihood*, not probability. And, unlike probability, we have no theory on how to use the likelihood as a weight for averaging, thus avoiding the paradox. Here is the point where probability-based reasoning fails: when we use probability as a weight for averaging.

Without the averages to compare the options, there is no rational basis to prefer to exchange the envelope. The likelihood satisfies the psychological need to attach some uncertainty to single events, something denied by frequentists, while avoiding the probability-induced paradox.

To get the *extended likelihood* explanation, first define X to be the content of the other/unchosen envelope, which must be either $2Y$ or $Y/2$. We can construct a binary random variable

$$U \equiv I(X = 2Y).$$

Once a choice is made, the values x, y and u are realized but remain unknown until both envelopes are opened. By the definition in Section 12.1, for the realized random parameter u, we have the extended likelihood

$$L_e(u = 1) = \Pr(X = 2Y) = 0.5,$$

so we have $x = 2y$ or $y/2$ with 50-50 extended-likelihood, not probability. This avoids the paradox. What's interesting in this solution is that we don't need to specify any fixed parameter, so it's theoretically the simplest explanation.

However, there is an equivalent and more intuitive language in terms of confidence. Recall from Section 12.3 that confidence is an extended likelihood, so equivalence is to be expected. For completeness, we will explain the confidence idea again. Firstly, it's clear that by choosing an envelope at random, we have the probability $\Pr(Y = \theta) = 0.5$. Now, reverse the inside argument of the probability and ponder for a moment:

$$\Pr(\theta = Y) = \Pr(\theta \in \{Y\}) = 0.5.$$

This is the probability of a correct guess of what θ is. To be more suggestive: $\{Y\}$ is a 50%-confidence guess of θ. *This means, once y is observed, we have 50% confidence that θ is equal to y or $y/2$.*

Consequently, the amount $X = x$ in the unchosen envelope is either $2y$ or $y/2$ with 50% confidence, not probability. We could also come to this result directly by interpreting the probability $\Pr(X = 2Y) = 0.5$ as a coverage probability. As before, the paradox is avoided because, again, we have yet no theory on how to take an average using confidence as weight. But we shall have such a theory in Section 19.5.

The frequentist explanation

The first type of frequentist thinking is a bit nihilistic: It simply declares that, given $y = 100$, the other envelope has either 50 (if $\theta = 50$) or 200 (if $\theta = 100$) *with probability one. There is no randomness.* So, stating $P(X = 200|y = 100) = P(X = 50|y = 100) = 0.5$ and using them to compute the expected value $E(X|y = 100)$ is wrong. Although technically true, the explanation is not helpful, since we are still in a state of uncertainty.

As discussed above, if you do not exchange or always exchange, your expected winning is $3\theta/2$. It is instructive to discuss a randomized strategy (Ross, 1994) that leads to better expected returns:

1. Generate a random variate $U = u$ with *any* strictly positive density $f(u)$ for $u > 0$.
2. Take one envelope at random and observe the amount $Y = y$.
3. Compare y and u. If $y < u$ (thus imagine that y is small), then exchange the envelope; otherwise, keep it.

We show in Appendix 1 of this chapter the proof that the expected winning of such a randomized strategy *is strictly greater* than $3\theta/2$. How much do we gain from using the randomized strategy? It obviously depends on which random variable U is used and its relation to the true θ. If we use the exponential variate, we obtain a maximum gain of around 8%, achieved when the exponential mean is close to the true θ.

Before you think that this frequentist strategy 'resolves' the paradox, let's imagine, however, how it can be applied in practice. *First, remember* *that θ is assumed to be fixed.* If you're going to play just once, the long-term average is not a meaningful quantity; other paradoxes will arise if you take the average seriously. So, let's assume you will play repeated games. Alert readers will immediately realize that the above strategy is *not* applicable to a rational person playing repeated games. To see this, suppose that at the first draw you see the amount $y = 100$, then you should ask for an exchange, so you'd find out what θ is exactly (either 50 or 100); after that, it will be absurd to use any randomized strategy.

So, the randomized strategy is relevant only for a group of individuals acting independently, each playing once; there is no sharing of any information; and at the end, the winnings are divided out equally. The question is whether this scenario is a realistic model of social behaviour. If they play the game independently, would the winners feel any obligation to share their winnings with the losers? Imagine independent traders on the stock market or independent gamblers in the casino.

An interesting, though perhaps fanciful, implication of the randomized strategy is for the gambling house that plans to offer the exchange game to the public. What price should they charge customers to play the game? They can't naively think of the base winning $3\theta/2$ as the fair price, as they'd lose money if the customers apply the randomized strategy.

The Bayesian explanation

If θ is random with a *known distribution*, a Bayesian strategy can be devised to take advantage of randomness in θ. Randomness of θ implies that the game is repeated with different values of θ each time. However, the setting we're interested in is when the game is played just once with an unknown fixed θ.

As the parameter θ is not an objective random variable, the prior distribution just reflects our ignorance (Chapter 9), as we know nothing about θ. Our intuitive reasoning that leads to the paradox corresponds to having the posterior probability

$$p(\theta|y) = 0.5, \quad \text{for } \theta = y/2, y.$$

Blachman et al. (1996) found that the posterior is implied by Jeffreys's invariant prior $1/\theta$ (Section 9.3). The proof, which is rather technical, is shown in this chapter's appendix. The question is whether the Bayesian explanation is a reasonable model for our intuitive mind, which arrives immediately at the 50-50 posterior probability.

The Bayesian explanation puts the blame on the improper prior as the cause of the paradox. That is, the paradox can be avoided if we use a proper prior. This means that, in contrast to the likelihood or confidence explanation, the Bayesian also denies the psychological sense of equal uncertainty.

19.4 Actual vs Counterfactual Envelopes

Another frequentist explanation of the exchange paradox is that, once you pick an envelope containing, say, $y =$ \$100, there is only *one further envelope* containing either \$200 or \$50, not two. These two values are counterfactual options, not actual sample-space values for probabilities, so we do not have a proper probability and cannot take the expected value.

Now let's change the counterfactual aspect and see what happens: After you choose and open one envelope to reveal $y =$ \$100, you will be shown *two actual* (closed) envelopes containing \$200 and \$50. Now, there should be no issues, as you're facing a bona-fide probability. Would you exchange your envelope with one of the other two? Now, it seems logical and uncontroversial to exchange, since you can calculate the expected value \$125 using real probability. This argument holds for any value of y, so exchanging is better even *before* opening the envelope.

By providing actual envelopes, we have indeed changed the overall expected winning. Conditional on $Y = y$, the winning is a proper random variable with an expected value y for staying put with the original envelope, and expected value $5y/4$ for exchanging. Since Y itself is random with an expected value $3\theta/2$, the overall expected values are $3\theta/2$ and $15\theta/8$ for the two strategies, respectively. So it seems there is no paradox that exchanging an unopened envelope is always better than staying put. But, before getting too comfortable, see this:

> **Factual reverse exchange paradox:** Instead of immediately exchanging your initial, unopened envelope, your random choice out of the other two envelopes is opened, and it reveals $y =$ \$100. Would you

> still be interested in exchanging? You're thinking: 'wait a minute, now I know my envelope contains either \$200 or \$50, giving an average of \$125. So it seems my initial choice is ok after all!' But this is again absurd, since we just discussed that, in this setting, exchanging is always better than staying put.

Thus the reverse version survives even when the original exchange paradox no longer holds. As before, the paradox occurs due to treating uncertainty as probability. Let's think of the possible values: if your envelope has \$200, then the other two must contain \$100 and \$400 and you happen to see the \$100. On the other hand, if your envelope has \$50, then the other two must contain \$25 and \$100 and again you happen to see \$100. Therefore, there are *four* possible values, whereas in reality there are only *three* envelopes. So, it's clear that \$200 and \$50 are counterfactual and we cannot attach probability to them. But, with a similar argument as before, what we have is 50% confidence rather than probability.

19.5 Risk Aversion and Confidence-Weighted Utility

We have previously discussed that the exchange paradox arises before we open the envelope. But after the envelope is opened, it is actually not obvious if we still have a paradox, since in the magnified exchange paradox we could be tempted to exchange. How do we explain this?

According to VNM's classic theorem (Section 7.4), one acts rationally by optimizing one's expected utility. Suppose we bet on one of k possible random outcomes with probabilities p_i for $i = 1, \ldots, k$, with respective winnings $w_1, \ldots, w_k$. The expected utility of the bet is

$$\mathcal{U} \equiv \sum_i p_i u(w_i), \tag{19.1}$$

where the $u(\cdot)$ is the utility function. By definition, $u(w)$ is the utility of the sure amount w; this sure amount is as good as the amount you have. If we have linear utility $u(w) = w$, then we get the standard expected value of the winning.

The utility of money can be thought of as the value of money. Empirically, for most people, the value of money does not increase linearly. The joy of winning \$1000 when you're poor is greater than the joy of winning the same amount when you're rich. In economics, this is known as the law of diminishing returns. This means that the shape of the utility function is typically concave. Let's look at an example:

> **Simple bet:** Would you rather receive a sure \$100, or an option to win either \$200 or \$50 with a fair flip of a coin? Now, the coin flip goes straight into classical probability, so you can legitimately compute the expected value of the bet as (200+50)/2=\$125. So you should be happy to take the bet option, right? In fact, no, many people would rather keep the sure \$100 than take the bet and risk a loss.
>
> This is called *risk aversion.* The expected utility of the bet is $\mathcal{U} = \{u(50) + u(200)\}/2$. Risk-averse behaviour is connected to the concave shape of the utility function, where higher gains are more heavily discounted, e.g., say $u(50) \approx 50$ but $u(200)$ could be much less than 200, so that the average utility is less than $u(100)$. As a result, a 'raw' and uncertain win of \$125 has lower attraction than a sure win of \$100. It is the betting application of the folk wisdom 'a bird in the hand is worth two in the bush.'

The shape of one's utility function is subjective, reflecting different levels of risk tolerance or aversion among people. Therefore, even when faced with the same situation, different individuals may arrive at different judgements and decisions. If not considered properly, the subjective nature of the utility function may lead to seemingly irrational human behaviour.

Remarkably, virtually all books on probability theory would say that taking the bet here is the logical thing to do. This probability-based reasoning is based on the implicit assumption that the offer will be *repeated indefinitely*, so that the expected value becomes a meaningful long-term average. Indeed, if it is stated explicitly that the bet would be repeated a large number of times, it would of course be irrational (or at least self-defeating) to refuse the bet. (The utility function would, of course, still work for the repeated bets. Technically, when a bet is repeated, the amount of money involved is larger, so we'll be in a different part of the utility function.)

Hence, the unstated assumption in the simple bet example is that the bet is played once, so it is a single-decision situation. As amusingly told by Ellenberg (2014, p. 198), '[in a single game] the expected value is not the value you expect!' This may explain why people intuitively do not simply use the standard expected value in single-decision problems.

Confidence-weighted utility

A final piece that we need to solve all the paradoxes in this chapter is the concept of confidence-weighted utility. Let's say we're betting on the random outcome $V = v_i$ with probability p_i for $i = 1, \ldots, k$. The classical definition of the expected utility (19.1) uses these probabilities. However, remember that expected utility can be used in single-decision problems. In such cases, we are dealing with a specific outcome $V = v$, which can be assumed to be *realized but still unknown* – so that the bet is still meaningful. Thus, V can be treated as an unknown random parameter, with extended likelihood

$$L_e(v = v_i) = \Pr(V = v_i) = p_i.$$

This means that we interpret the expected utility (19.1) as an average with the extended likelihood as a weight. However, as before, we can use an equivalent but more intuitive language of confidence. Each p_i is the confidence in $v = v_i$, so that (19.1) is a confidence-weighted utility.

To emphasize, this is not just a trivial restatement of the definition. Theoretically, objective probability applies only to sample-space values, not counterfactuals; in contrast, as highlighted in the exchange paradox, likelihood or confidence can apply to counterfactuals. We have shown that in the original exchange paradox, we have 50-50 confidence rather than 50-50 probability. Previously, we had no theory on how to compute an average using confidence as a weight. Our discussion here implies that *what we can compute is the confidence-weighted utility.* This justifies the temptation to exchange the envelope in the magnified exchange paradox.

To get a more intuitive appreciation, let's compare the following three situations as a single-decision problem; the numbers are taken from the magnified exchange paradox, where exchanging is tempting. Are the situations logically and psychologically equivalent so that you will make the same decision in all situations?

Situation A: Would you rather get a sure \$100, or a chance to win either \$10,000 or \$1 with a fair flip of a coin? In this situation, we can imagine two actual envelopes containing \$10,000 and \$1, so we are now in the realm of proper probability. Would you take the bet? Unless you're super risk-averse, you will likely take the bet, so let's say you will. This means your $u(100)$ is smaller than $\{u(1) + u(10,000)\}/2$.

Situation B: In Situation A we do not say anything about the coin flip, so you may have assumed that it has not been performed and the use of probability is mathematically correct. Suppose that you are in the same situation as A, but now you are told that *the flip had been done* (say by someone you trust, so there is no issue of fairness or randomness), and the result had been used to select one of the two unopened envelopes for you. Hence, you only face a single, unopened envelope, containing either \$10,000 or \$1. Would you exchange your sure \$100 for the uncertain amount in the envelope?

Situation C: Assume you are in the original magnified exchange paradox setup. You open your envelope and see $y = \$100$; there is only one other envelope, containing either \$10,000 or \$1 with equal confidence. Would you be tempted to exchange?

First, compare A and B: Is your sense of uncertainty affected by whether the coin has been tossed in the past or is to be tossed in the future? For many decades, some US publishing houses – such as Publishers Clearing House and American Family Publishers – attracted subscribers by mailing sweepstakes envelopes that prominently declared, 'You May Have Won One Million Dollars!' All you have to do is return an attached form that contains a pre-selected 'winning number,' thereby entering the sweepstakes or lottery. The pre-selected 'winning number' may have confused a lot of old retirees into subscribing to more magazines than they could ever read, but do you believe that you have a better chance at winning because the winning number has been pre-selected compared to if the number is to be selected in the future? For most of us, A and B are logically equivalent, so we would feel the same and make the same decision. What makes betting on a random event a bet is your ignorance of the outcome, not the time factor.

The expected utility in A is naturally computed using probability. As discussed before, once a coin is tossed but the outcome is still unknown, the sense of uncertainty is captured by confidence. So, in B we have a confidence-weighted utility, which must then be equal to the expected utility in A.

What about B vs C? Physically, you are in the same position: facing an unopened envelope containing a fixed amount of either \$10,000 or \$1, while having a guaranteed \$100 at hand. In both situations, the uncertainty is captured by 50-50 confidence, not 50-50 probability. This means that in C we can also compute the confidence-weighted utility. Hence,

B and C are logically indistinguishable, so you would be compelled to make the same decision.

Put together, if we accept that Situations A, B and C are logically equivalent, it explains why one would be tempted to exchange in the magnified exchange paradox: because the confidence-weighted utility would indicate it is advantageous to do so.

19.6 Discussion

The exchange paradox shows a surprising richness that has allowed us to learn more about the nature of probability-based and likelihood-based reasoning, particularly for single events. The intuitive use of subjective probability can lead to paradoxes, which can be traced to the simple averaging of gains and losses. The keys to explaining the paradoxes involve (i) recognition of the role of confidence as a measure of uncertainty, and (ii) using confidence as a weight for averaging the utility function.

The utility theory explains why, in the magnified exchange paradox, it's legitimate to be tempted to exchange after opening the envelope. But note that it also explains why it is not rational to exchange before opening the envelope. Without knowing the actual value of y, the comparison of $\{u(y/2) + u(2y)\}/2$ versus $u(y)$ is, of course, indeterminate. The subjective utility function needs actual values: We may feel differently if $y = \$100$ or if $y = \$10,000$.

The existence of a paradox is a warning that there is something incomplete, if not wrong, in our reasoning. What's wrong in the exchange paradox is perhaps the lack of appreciation for the stark logical difference in the application of probability for single vs repeated decisions. The standard expected value is relevant in the latter, but the former requires utility theory. And what's incomplete is the knowledge of the extended likelihood, its close connection to the concept of confidence as measures of uncertainty (Chapter 12) and its application in the computation of the confidence-weighted utility.

Appendix 1: Randomized Strategy

Here is the strategy in which the decision to exchange is determined by a random number. The steps are as follows:

1. Generate $u \sim f(u)$, where $f(\cdot)$ is a known probability density for $u > 0$.

2. Take one envelope and observe $Y = y$.

3. Compare: if $y > u$ keep the envelope, otherwise exchange it.

Now we show that the expected winning $EW(Y) > 3\theta/2$. Fixing $U = u$, we have

$$\begin{aligned} W(y) &= yI(y > u) + 2yI(y < u), \\ &= \begin{cases} \theta I(\theta > u) + 2\theta I(\theta < u), & \text{if } y = \theta \\ \theta I(2\theta < u) + 2\theta I(2\theta > u), & \text{if } y = 2\theta, \end{cases} \end{aligned}$$

where $I(\cdot)$ is the indicator function. So,

$$\begin{aligned} E(W|u) &= \frac{3\theta}{2} + \frac{\theta}{2}\{I(u < 2\theta) - I(u < \theta)\}, \\ EW &= \frac{3\theta}{2} + \frac{\theta}{2}\{P(U < 2\theta) - P(U < \theta)\}, \\ &> \frac{3\theta}{2}. \end{aligned}$$

Suppose we use the exponential variate with mean μ, so that $P(U < \theta) = 1 - e^{-\theta/\mu}$. Figure 19.1 shows the ratio of the expected winning $EW/(3\theta/2)$ as a function of μ/θ. A maximum gain of around 8% is achieved when μ is close to the true θ.

Appendix 2: Bayesian Solution

Let $g(\theta)$ be the prior density of θ. The random variable Y is a transformed θ from two sources: $Y = 2\theta$ and $Y = \theta$ with probability 1/2

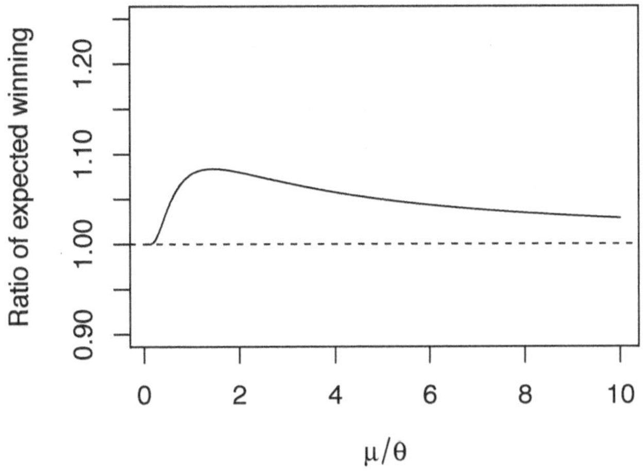

FIGURE 19.1
Ratio of the expected winning of the randomized strategy vs the base strategy of no exchange as a function of the ratio of exponential mean μ over the true θ.

each, so

$$p(y) = \frac{1}{2}\sum_{\theta} g(\theta)|\partial\theta/\partial y| = g(y/2)/4 + g(y)/2,$$

where the sum is over $\theta = y/2$ and y, and the Jacobian terms $|\partial\theta/\partial y|$ are 1/2 and 1 respectively. So, the posterior probability is

$$p(\theta|y) = \frac{g(\theta)|\partial\theta/\partial y|/2}{p(y)}, \text{ for } \theta = y/2, y.$$

If we set $p(\theta|y) = 1/2$, we get the constraint

$$g(\theta/2) = 2g(\theta),$$

which is satisfied by Jeffreys's invariant prior $g(\theta) \propto 1/\theta$.

For comparison, here the confidence and the likelihood are proportional, so the implied prior in the confidence solution is

$$c_0(\theta; y) \propto \frac{c(\theta; y)}{L(\theta; y)} = 1 \text{ for } \theta \in \{y/2, y\},$$

i.e., we have a data-dependent uniform implied prior.

Bibliography

Allais, M. 1953. Le comportement de l'homme rationnel devant le risque—Critique des postulats de l'Ecole Americaine. *Econometrica*, **21**: 503–546.

Armitage, P. (1961). Contribution to the discussion of 'Consistency in statistical inference and decision,' by C.A.B. Smith. *Journal of the Royal Statistical Society*, B, **23**, 1–37.

Arrow, K.J. (1950). A Difficulty in the Concept of Social Welfare. *The Journal of Political Economy*, **58**, 328–346.

Arrow, K.Jånd Debreu, G. (1954). Existence of an Equilibrium for a Competitive Economy. *Econometrica*, **22**, 265–290.

Balch, M.S., Martin, R. and Ferson, S. (2019). Satellite conjunction analysis and the false confidence theorem. *Proceedings of the Royal Society A*, **475**, 20180565.

Barnard, G. A. (1947), Significance tests for 2×2 tables. *Biometrika*, **34**, 123–138.

Barnard, G. A. (1949). Statistical inference. *Journal of the Royal Statistical Society,* B, **11**, 115–139.

Basu, D. (1975). Statistical information and likelihood. *Sankhya: The Indian Journal of Statistics,* A, **37**, 1–71.

Bayarri, M.J., DeGroot, M.H. and Kadane, J.B. (1988). What is the likelihood function? In *Statistical Decision Theory and Related Topics IV*, Vol. 1. New York: Springer.

Berger, J.O. and Wolpert, R. (1988). *The Likelihood Principle.* Hayward: Institute of Mathematical Statistics Monograph Series.

Bernardo. J.M. (1979). Reference posterior distributions for Bayesian inference (with discussion). *Journal of the Royal Statistical Society: Series B (Methodological)*, **41**, 113–147.

Bernardo, J.M. and Smith, A.F.M. (1994). *Bayesian Theory.* Chichester: Wiley.

Bertrand, J. (1889). *Calcul des probabilités.* Gauthier-Villars, pp. 5–6.

Binmore, K. (2009). *Rational Decisions.* Princeton: Princeton University Press.

Birnbaum, A. (1962). On the foundation of statistical inference. *Journal of the American Statistical Association*, **57**, 269–326.

Bjørnstad, J.F. (1996). On the generalization of the likelihood function and likelihood principle. *Journal of the American Statistical Association*, **91**, 791–806.

Blachman, N.M., Christensen, R. and Utts, J.M. (1996). Letter with corrections to the original article by Christensen, R., and Utts, J.M. (1992). *The American Statistician*, **50**, 98–99.

Butler, R.W. (1986). Predictive likelihood inference with applications. *Journal of the Royal Statistical Society: Series B (Methodological)*, **48**, 1–38.

Camerer, C. and Weber, M. (1992). Recent developments in deasuring preferences: uncertainty and ambiguity. *Journal of Risk and Uncertainty*, **5**, 325–370.

Carnap, R. (1950). *Logical Foundations of Probability.* Chicago: University of Chicago Press.

Choi, S. Blume, J.D. and Dupont, W.D. (2013). Elucidating the foundations of statistical inference with 2×2 tables. *PLoS ONE*, **10**, e0121263.

Christensen, R. and Utts, J.M. (1992). Bayesian resolution of the exchange paradox. *The American Statistician*, **46**, 274–276.

Cohen, L.J. (1977). *The Probable and the Provable.* Oxford: Clarendon Press.

Cox, D.R. and Hinkley, D.V. (1979). *Theoretical Statistics.* London: Chapman and Hall.

Crupi, V., Fitelson, B. and Tentori, K. (2007). Probability, confirmation, and the conjunction fallacy. *Thinking and Reasoning*, **14**, 182–199.

Dale, A.I. (1999). *A History of Inverse Probability: From Thomas Bayes to Karl Pearson.* 2nd ed. New York: Springer.

Daston, L. (1988). *Classical Probability in the Enlightenment.* Princeton: Princeton University Press.

Dawid, A.P., Stone, M. and Zidek, J.V. (1973). Marginalization paradoxes in Bayesian and structural inference (with discussion). *Journal of the Royal Statistical Society: Series B (Methodological)*, **35**, 189–233.

Delgado-Rodrıguez, M. and Llorca, J. (2004). Biases. *Journal of Epidemiology and Community Health*, **58**, 635–641.

Descartes, R. (1641). *Meditations on First Philosophy.* Translation by M. Moriarty in 2008, Oxford University Press.

de Finetti, B. (1931). On the subjective meaning of probability. English translation in de Finetti (1993), *Induction and Probability.* Bologna: Clueb, pp. 291–321.

de Finetti, B. (1974). *Theory of Probability* (Vol. I). London: Wiley.

Doidge, N. (2008). *The Brain that Changes Itself.* London: Penguin Books.

Donnelly, P. and Friedman, R.D. (1999). DNA database searches and the legal consumption of scientific evidence. *Mich. Law Rev.*, **97**, 931–984.

Drory, A. (2015). Failure and uses of Jaynes' principle of transformation groups. *Foundations of Physics*, **45**, 439–460.

Easwaran, K. (2013). Why countable additivity? *Thought*, **2**, 53–61.

Edwards, A.W.F. (1972). *Likelihood.* Cambridge: Cambridge University Press. Expanded edition in 1992, Baltimore: Johns Hopkins University Press.

Ellenberg, J. (2014). *How Not To Be Wrong: The Power of Mathematical Thinking.* New York: Penguin Books.

Ellsberg, D. (1961). Risk, ambiguity, and the Savage axioms. *Quarterly Journal of Economics*, **75**, 643–669.

Empiricus, S. (1933). *Outlines of Pyrrhonism.* Translated by R.G. Bury (Loeb edn). London: W. Heinemann, p. 283.

Fisher, R.A. (1921). On the 'probable error' of a coefficient of correlation deduced from a small sample. *Metron*, **1**, 2–32.

Fisher, R. A. (1922). On the mathematical foundations of theoretical statistics. *Philosophical Transactions of the Royal Society A: Mathematical, Physical and Engineering Sciences*, **222**, 309–368.

Fisher, R.A. (1930). Inverse probability. *Proceedings of the Cambridge Philosophical Society*, **26**, 528–535.

Fisher, R.A. (1933). The concepts of inverse probability and fiducial probability referring to unknown parameters. *Proceedings of the Royal Society A*, **139**, 343–348.

Fisher, R.A. (1935). *The Design of Experiments*, New York: Hafner.

Fisher, R.A. (1955). Statistical Methods and Scientific Induction. *Journal of the Royal Statistical Society,* B, **17**, 69–78.

Fisher, R.A. (1958). The nature of probability. *Centennial Review*, **2**, 261–274.

Fisher, R.A. (1959). *Statistical methods and scientific inference.* 2nd Edition. Edinburgh: Oliver & Boyd. Posthumous 3rd edition, 1973, New York: Hafner.

Gardner, M. (1959). Mathematical games: Problems involving questions of probability and ambiguity. *Scientific American*, **201**, 174–182.

Gardner, M. (1982). *Aha! Gotcha: Paradoxes to Puzzle and Delight.* San Francisco: W H Freeman & Co.

Gardner-Medwin, T. (2005). What probability should a jury address? *Significance*, **2**, 9–12.

Gilboa, I. and Schmeidler, D. (1993). Updating ambiguous beliefs. *Journal of Economic Theory*, **59**, 33–49.

Gill, R. (2011). The Monty Hall problem is not a probability puzzle (It's a challenge in mathematical modelling). *Statistica Neerlandica*, **65**, 58–71.

Gillies, D. (2000). *Philosophical Theories of Probability.* Routledge: London.

Gödel, K. (1931). Über formal unentscheidbare Sätze der Principia Mathematica und verwandter Systeme I. *Monatshefte Für Mathematik und Physik*, 38, 173–198.

Good, I.J. (1960). The Paradox of Confirmation. *British Journal for the Philosophy of Science*, **11**, 145–149.

Good, I.J. (1961). The Paradox of Confirmation (II). *British Journal for the Philosophy of Science*, **12**, 63–64.

Gould, S.J. (1988). The Streak of Streaks. *The New York Review of Books*, **35** (13).

Gweon H., Tenenbaum J.B. and Schulz L.E. (2010). Infants consider both the sample and the sampling process in inductive generalization. *Proceedings of the National Academy of Sciences*, **107**, 9066–9071.

Harari, Y.N. (2015). *Sapiens: A Brief History of Humankind.* New York: Harper Perennial.

Hartmann, L. (2020). Savage's P3 is redundant. *Econometrica*, **88**, 203–205.

Heisenberg, W. (1958). *Physics and Philosophy.* Allen and Unwin: London.

Hempel, C.G. (1945a). On the nature of mathematical truth. *The American Mathematical Monthly*, **52**, 543–556.

Hempel, C.G. (1945b). Studies in the logic of confirmation I. *Mind*, **54**, 1–26.

Henderson, C.R., Kempthorne, O., Searle, S.R. and Von Krosigk, C.M. (1959). The estimation of environmental and genetic trends from records subject to culling. *Biometrics*, **15**, 192–218.

Hertwig, R. and Gigerenzer, G. (1999). The 'conjunction fallacy' revisited: How intelligent inferences look like reasoning errors. *Journal of Behavioral Decision Making*, **12**, 275–305.

Hicks, J. (1980). *Causality in Economics.* Canberra: ANU Press.

Hintikka, J. (2004). A fallacious fallacy? *Synthese*, **140**, *Knowledge and Decision: Essays on Isaac Levi*, pp. 25–35.

Howson, C. and Urbach, P. (1989). *Scientific Reasoning, the Bayesian Approach.* Chicago and La Salle: Open Court.

Hume, D. (1748). *An Enquiry Concerning Human Understanding.* 2006 Edition. Oxford: Clarendon Press.

Jaynes, E.T. (1973). The well-posed problem. *Foundations of Physics*, **3**, 477–493.

Jaynes, E.T. (2003). *Probability Theory: The Logic of Science.* Cambridge: Cambridge University Press.

Jeffreys, H. (1961). *Theory of Probability.* Oxford: Oxford University Press.

Kahneman, D. (2011). *Thinking, Fast and Slow.* New York: Farrar, Straus and Giroux.

Kahneman, D. and Tversky A. (1979). Prospect Theory: An Analysis of Decision under Risk. *Econometrica*, **47**: 263–292.

Kant, I. (1781). *Critique of Pure Reason.* Translation by M. Weigelt in 2003, Penguin Classics.

Kolmogorov, A.N. (1933). *Grundbegriffe der Wahrscheinlichkeitrechnung.* English translation by N. Morrison, *Foundations of the Theory of Probability.* New York: Chelsea, 1956.

Koopman, B.O. (1940). The bases of probability. *Bulletin of the American Mathematical Society*, **46**, 763–774.

Kuhn, T. (1962). *The Structure of Scientific Revolutions.* Chicago: Chicago University Press.

Kyburg, H.E. (1961). *Probability and the Logic of Rational Belief.* Middletown, CT: Wesleyan University Press.

Laplace, P.S. (1814). *A Philosophical Essay on Probabilities.* English translation by F.W. Truscott and F.L. Emory in 1902 from the 6th French edition.

Lee, H. and Lee, Y. (2023a). H-likelihood approach to deep neural networks with temporal-spatial random effects for high-cardinality categorical features. *Proceedings of the 40th International Conference on Machine Learning*, PMLR **202**: 18974–18987.

Lee, H. and Lee, Y. (2023b). Point mass in the confidence distribution: Is it a drawback or an advantage? *arXiv preprint arXiv: 2310.09960.*

Lee, H. and Lee, Y. (2023c). On the statistical foundations of h-likelihood for unobserved random variables. *arXiv preprint arXiv: 2310.09955.*

Lee, Y. (2020). Resolving the induction problem: Can we state with complete confidence via induction that the sun rises forever? *arXiv preprint arXiv: 2001.04110.*

Lee, Y. and Bjørnstad, J.F. (2013). Extended likelihood approach to large-scale multiple testing. *Journal of the Royal Statistical Society: Series B (Methodological)*, **75**, 553–575.

Lee, Y. and Nelder, J.A. (1996). Hierarchical generalized linear models (with discussion). *Journal of the Royal Statistical Society: Series B (Methodological)*, **58**, 619–678.

Lee Y., Nelder J.A. and Pawitan, Y. (2017). *Generalized Linear Models with Random Effects: Unified Analysis via H-likelihood.* 2nd Edition. Chapman and Hall/CRC.

Lenzen, W. (2017). Leibniz's ontological proof of the existence of God and the problem of impossible objects. *Logica Universalis*, **11**, 85–104.

Lewis, D. (1980). A subjectivist's guide to objective chance. In W.L. Harper, R. Stalnaker and G. Pearce (Eds.), *Ifs*, Springer, pp. 267–297.

Lindley, D.V. (2000). The philosophy of statistics. *Journal of the Royal Statistical Society: Series D (The Statistician)*, **49**, 293–337.

Lipton, P. (1993). *Inference to the Best Explanation.* London: Routledge.

Luce, R. D. and Raiffa, H. (1957). *Games and Decisions: Introduction and Critical Survey.* New York: Wiley.

Ludbrook, J. (2013). Analysing 2×2 contingency tables: Which test is best? *Clinical and Experimental Pharmacology and Physiology*, **40**, 177–180.

Mayo, D.G. (2014). On the Birnbaum argument for the strong likelihood principle. *Statistical Science*, **29**, 227–239.

McCall, S. and Armstrong, D.M. (1989). God's lottery. *Analysis*, **49**, 223–224.

Moro, R. (2009). On the nature of the conjunction fallacy. *Synthese*, **171**, 1–24.

National Research Council. (1996). *The Evaluation of Forensic DNA Evidence.* Washington, DC: The National Academies Press. `https://doi.org/10.17226/5141`.

Neyman, J. and Pearson, E.S. (1933). On the problem of the most efficient tests of statistical hypotheses. *Philosophical Transactions of the Royal Society A: Mathematical, Physical and Engineering Sciences*, **231**, 289–337.

Neyman, J. (1937). Outline of a theory of statistical estimation based on the classical theory of probability. *Philosophical Transactions of the Royal Society A: Mathematical, Physical and Engineering Sciences*, **236**, 333–380.

Osherson, D.N., Smith, E.E., Wilkie, O., López, A. and Shafir, E. (1990). Category-based induction. *Psychological Review*, **97**, 185–200.

Pawitan, Y. (2001). *In All Likelihood: Statistical Modelling and Inference Using Likelihood.* Oxford: Clarendon Press.

Pawitan, Y. (2004). Likelihood perspectives in the consensus and controversies of statistical modelling and inference. In Adams N. M. et al. (2004). *Methods and Models in Statistics: In Honour of Professor John Nelder, FRS.* London: Imperial College Press.

Pawitan, Y. and Lee, Y. (2017). Wallet game: Probability, likelihood and extended likelihood. *The American Statistician*, **71**, 120–122.

Pawitan, Y. and Lee, Y. (2021). Confidence as likelihood. *Statistical Science*, **36**, 509–517.

Pawitan, Y., Lee, H. and Lee, Y. (2023). Epistemic confidence in the observed confidence interval. *Scandinavian Journal of Statistics*. Published online 12 April 2023, `https://doi.org/10.1111/sjos.12654`

Peirce, C.S. (1867). On the natural classification of arguments. *Proceedings of the American Academy of Arts and Sciences*, **7**, 261–287.

Peirce, C.S. (1883). A theory of probable inference. In Peirce, Charles S. (ed.). *Studies in Logic by Members of the Johns Hopkins University*. Boston, MA.

Pitman, E.J.G. (1937). Significance tests which may be applied to samples from any population. *Royal Statistical Society Supplement*, **4**, 119–130 and 225–232 (parts I and II).

Polya, G. (1954). *Mathematics and Plausible Reasoning Volume I: Induction and Analogy in Mathematics*. Princeton: Princeton University Press.

Popper, K.R. (1957). The propensity interpretation of the calculus of probability, and the quantum theory. In S. Körner (ed.) *Observation and Interpretation*. Proceedings of the Ninth Symposium of the Colston Research Society, University of Bristol, pp. 65–70.

Popper, K.R. (1959). *The Logic of Scientific Discovery*. 6th revised impression of the 1959 English translation.

Popper, K.R. (1990). *A World of Propensities*. Bristol: Thoemmes Press.

Raiffa, H. (1961). Risk, ambiguity, and the Savage axioms: Comment. *Quarterly Journal of Economics*, **75**, 690–94.

Ramsey, F. (1931). Truth and probability. In F. Ramsey (Ed.), *Foundations of Mathematics and Other Logical Essays*. New York: Harcourt.

Reichenbach, H. (1949). *The Theory of Probability*. Berkeley: University of California Press.

Royall, R. (1997). *Statistical Evidence: A Likelihood Paradigm.* New York: Chapman & Hall.

Roeder, K. (1994). DNA fingerprinting: A review of the controversy. *Statistical Science*, **9**, 222–247.

Ross, S. (1994). Comment to Christensen, R. and Utts, J. (1992). *The American Statistician*, **48**, 267.

Ross, S. (1976). Return, risk and arbitrage. In: I. Friend and J. Bicksler (eds.), *Studies in Risk and Return.* Cambridge, MA: Ballinger

Russell, B. (2016). *The Impact of Science on Society.* London: Routledge (originally published in 1952).

Ryder, J.M. (1981). Consequences of a simple extension of the Dutch Book argument. *British Journal for the Philosophy of Science*, **32**, 164–167.

Savage, L. (1972). *The Foundations of Statistics.* 2nd Edition. New York: Wiley.

Schervish, M., Seidenfeld, T. and Kadane, J. (1984). The extent of non-conglomerability of finitely additive probabilities. *Zeitschrift für Wahrscheinlichkeitstheorie und Verwandte Gebiete*, **66**, 205–226.

Schweder, T. (2018). Confidence is epistemic probability for empirical science. *Journal of Statistical Planning and Inference*, **195**, 116–125.

Schweder, T. and Hjort, N.L. (2016). *Confidence, Likelihood, Probability: Statistical Inference with Confidence Distributions.* Cambridge: Cambridge University Press.

Senn, S. (2003). *Dicing with Death: Chance, Risk and Health.* 1st Edition. Cambridge: Cambridge University Press.

Senn, S. (2009). Comment on Harold Jeffreys's *Theory of Probability Revisited. Statistical Science*, **24**, 185–186

Selvin, S. (1975a). A problem in probability. *The American Statistician*, **29**, 67–71.

Selvin, S. (1975b). On the Monty Hall problem. *The American Statistician*, **29**, 134.

Shafer, G. (1986). Savage revisited. *Statistical Science*, **1**, 463–485.

Shafir, E.B., Smith, E.E. and Osherson, D.N. (1990). Typicality and reasoning fallacies. *Memory and Cognition*, **8**, 229–239.

Sinn, H.W. (1980). A rehabilitation of the principle of insufficient reason. *The Quarterly Journal of Economics*, **94**, 493–506.

Soros, G. (1987). *The Alchemy of Finance*. Hoboken, New Jersey: Wiley.

Stefánsson, H.O., and Bradley, R. (2015). How valuable are chances? *Philosophy of Science*, **82**, 602–625.

Stein, C. (1959). An example of wide discrepancy between fiducial and confidence intervals. *The Annals of Mathematical Statistics*, **30**, 877–880.

Stigler, S.M. (1986). Laplace's 1774 memoir on inverse probability. *Statistical Science*, **1**, 359–363.

Stone, M. and Dawid, A.P. (1972). Un-Bayesian implications of improper Bayesian inference in routine statistical problems. *Biometrika*, **59**, 369–375.

Suissa, S. and Schuster, J. (1985). Exact unconditional sample sizes for the 2×2 binomial trial. *Journal of the Royal Statistical Society (B)*, **148**, 317–327.

Thaler, R.H. (2016). *Misbehaving: The Making of Behavioral Economics*. New York: W.W. Norton & Co.

Tversky, A. and Kahneman, D. (1983). Extensional versus intuitive reasoning: The conjunction fallacy in probability judgment. *Psychological Review*, **90**, 293–315.

Venn, J. (1876). *The Logic of Chance*. 2nd Edition. Macmillan & Co.

Villegas, C. (1964). On qualitative probability σ-algebras. *Annals of Mathematical Statistics*, **35**, 1787–1796.

von Mises, R. (1964). *Mathematical Theory of Probability and Statistics*. New York: Academic Press.

von Neumann, J., and Morgenstern, O. (1947). *The Theory of Games and Economic Behavior*. 2nd Edition. Princeton: Princeton University Press.

vos Savant, M. (1997). *The Power of Logical Thinking*. New York: St. Martin's Griffin.

Welch, B.L. and Peers, H.W. (1963). On formulae for confidence points based on integrals of weighted likelihoods. *Journal of the Royal Statistical Society: Series B (Methodological)*, **25**, 318–329.

Whitehead, A.N. and Russell, B. (1910). *Principia Mathematica*. Vol. I. Cambridge: University Press.

Williamson, J. (1999). Countable additivity and subjective probability. *The British Journal for the Philosophy of Science*, **50**, 401–416.

Index